GREENHOUSE

GREENHOUSE

The 200-Year Story
of Global Warming

Gale E. Christianson

Constable • London

To Rhonda

~

First published in the United States of America in 1999
by Walker Publishing Company, Inc.

First published in Great Britain 1999
by Constable and Company Limited
3 The Lanchesters, 162 Fulham Palace Road
London W6 9ER

The right of Gale E. Christianson to be identified as author of this work has been asserted
by him in accordance with the Copyright, Designs and Patents Act 1988

A CIP catalogue record for this book is available from the British Library

Printed in the United States of America

HOW CAM'ST THOU IN THIS PICKLE?

William Shakespeare, *The Tempest*

Contents

Preface   ix

Part One
THE TIME TRAVELERS
~

1. The Guillotine and the Bell Jar   3
2. The Cryptic Moth   13
3. "Endless and as Nothing"   24

Part Two
THE WORLD EATERS
~

4. "Quest for the Black Diamond"   39
5. Cleopatra's Needles   54
6. Vulcan's Anvil   66
7. The Phantom of the Open Hearth   75
8. "The Dynamo and the Virgin"   92

Part Three
THE DWELLERS IN THE CRYSTAL PALACE
~

9. Native Son   105
10. "Never a Man"   116
11. Threshold   135
12. A Tap on the Shoulder   145
13. Pendulum   158

14. A Death in the Amazon   172
15. The Climatic Flywheel   192
16. Cassandra's Listeners   210
17. Signs and Portents   222
18. Scenarios   235
19. Kyoto   254
   Coda   269

Bibliography   279
Acknowledgments   293
Index   295

Preface

One day, at the beginning of the geologic epoch called the Pleistocene, Earth's skies turned an ominous gray. The north wind rose and began churning the frigid air into keening gusts. Soon the first snowflakes descended, signaling the onset of what the nature writer Loren Eiseley termed "the angry winter." It barely stopped snowing for the next 2 million years, during which the planet was held in thrall by massive ice sheets that covered all of Antarctica, most of Europe, large expanses of North and South America, and lesser parts of Asia. The only sounds were the thunder of great avalanches and the gnashing of the advancing glaciers. The most recent of the great ice ages had arrived, triggered perhaps by the completion of the Milky Way's latest rotation, which occurs only once every 300 million years.

Scientists tell us that we remain citizens of the Ice Age. And many of them believe that it is only a matter of time before Earth's surface disappears once more beneath the blinding snows and mile-thick glaciers. When this will come to pass, no one can say for certain.

At present, climatologists are preoccupied by a more immediate concern than the next revolution of the galactic wheel. For much of the last century, Earth and its atmosphere have been heating up, a process that most, though not all, scientists believe is due to the massive consumption of fossil fuels—coal, oil, natural gas—triggered by the industrial revolution. What is more, global warming is accelerating. The 1970s were warmer than the 1960s; the 1980s were warmer than the 1970s; and the 1990s have been warmer still.

Global warming is not a newly discovered phenomenon.

The early debates over its cause were mostly conducted in scientific journals in technical language inaccessible to the public at large. This is no longer the case. In the span of little more than a week during the summer of 1998, the following headlines appeared in newspapers delivered at my doorstep. All but one made the front page: "July Breaks Worldwide Temperature Record," "Warmer, Wetter, Sicker: Linking Climate to Health," "Drought in Texas and Oklahoma Stunting Crops and Economies: Severity Is Reminiscent of the Dust Bowl Years," "Frogs Falling Silent Across USA," and "Religious Groups Mount a Campaign to Support Pact on Global Warming." Their collective message is disquieting to say the least.

Greenhouse is the biography of a scientific idea, the story of what global warming—or the so-called greenhouse effect—is and of how it came to be. The story begins nearly two centuries ago, with the natural philosopher Jean-Baptiste-Joseph Fourier, who came within a hair's breadth of being executed during the French Revolution. Fourier was the first to envision Earth as a giant greenhouse whose atmosphere traps the radiant heat from the Sun, warming the planet and giving life to every plant and animal inhabiting its surface, a sign to the Frenchman of nature's great benevolence.

Fourier and the other great scientific figures in part one were "The Time Travelers." Before them, Earth's origins and age were based on the chronology set forth in the Old Testament, and it was impossible to grasp the sweep of time or the great changes, both climatic and geologic, to which the planet has been subjected during the 4 billion years of its existence. Among Fourier's fellow voyagers were James Hutton, a lonely, contemplative Scotsman, and Sir Charles Lyell, an inveterate collector of butterflies. Together they championed the theory of uniformitarianism, which argues that the atmospheric and geologic forces currently at work are the same as those that operated in the past, paving the way for the modern science of geology. And for Charles Darwin as well. Lyell's friend and colleague, Darwin added time to the evolutionary scale the way Copernicus, Galileo, and Newton added distance to the stars. Yet Darwin would have rested much easier had he only known about a simple moth whose coloration was changing in response to the polluted skies of an industrializing England.

In a series of benumbing changes that bordered on the inexorable, the warp and woof of nature were being rewoven on the loom of industry. In part two, "The World Eaters," we encounter the inventors and the capitalists who wrested fossil fuels from the earth and used them to transform the planet: Richard Arkwright, the textile manufacturer who invented the factory system; ironmaster Abraham Darby of Coalbrookdale; Thomas Newcomen and James Watt, who perfected the steam engine; locomotive builders Richard Trevithick and George Stephenson; steel magnates Henry Bessemer and Andrew Carnegie; John D. Rockefeller, the baron of Standard Oil; Sir Isambard Kingdom Brunel, the nineteenth century's most colorful and renowned engineer; Model T builder Henry Ford. Towering hundreds of feet above it all and spewing their burden into the atmosphere twenty-four hours a day were Cleopatra's Needles, the massive smokestacks of brick and stone that became the preeminent symbol of the industrial age. So seeming complete was the victory of man over nature that the historian Henry Adams, while visiting the Hall of Dynamos at the Great Exposition in Paris in 1900, suddenly felt that the planet had been greatly diminished.

In 1896, three-quarters of a century after Fourier published his all but forgotten article, the Swedish chemist Svante Arrhenius returned to the subject. The future Nobel laureate conjectured that industrial pollutants, most particularly carbon dioxide, were accumulating in Earth's atmosphere. If this gaseous buildup continued, temperatures would gradually rise, although Arrhenius believed it likely that the world's supply of coal and other carbon fuels would be exhausted long before global warming could have any appreciable effect.

As we have recently learned, only a few degrees separate a warm planet from one shrouded in ice, a tale told in the once mysterious disappearance of the Anasazi from the American Southwest, the parallel demise of the Vikings of Greenland, and the scattering of the "Okies" during the Dust Bowl of the Great Depression. Not until the late 1970s would scientists, armed with data indicating that the planet was warming more rapidly than Arrhenius theorized, sound the alarm. In part three, "The Dwellers in the Crystal Palace," we trace the major discoveries and follow the debates, both scientific and public, from the early twentieth century through the 1997 United Nations Conference on Climate Change in Kyoto and beyond.

Among the individuals who figure prominently in these developments is George Callendar, an obscure English coal engineer who, standing alone for three decades, clung doggedly to Arrhenius's hypothesis that humans are indeed capable of altering the wind and weather. A citizen of London, Callendar watched in disbelief as killer fogs gave way to killer smogs, the deadliest of which claimed more than 4,000 lives in 1952. On Callendar's heels came Charles Keeling, a gifted and rebellious chemist whose obsession with trapping air in glass flasks accidentally led to the discovery that carbon dioxide levels in the atmosphere are rising more rapidly than anyone had previously theorized, and that they seem to be directly related to the greenhouse effect. Another brilliant chemist, Thomas Midgley Jr., invented chlorofluorocarbons for use in refrigerators and air-conditioning units, little suspecting that his chemicals would rend the ozone layer above Antarctica, exposing the world's plants and animals to the dangers of ultraviolet radiation. It would take Joseph Farman, a member of the British Antarctic Survey, twenty-seven years on the bone-numbing continent to alert the world to Midgley's potentially catastrophic blunder.

In the rain forests of the world dwell most of Earth's plant, animal, and insect species, the majority of which are not yet classified by scientists. Locked within the massive canopy are countless billions of tons of carbon dioxide. It is being returned to the atmosphere at a record rate as the result of intentional burning and rampant logging, which, when opposed by environmentalists, has often resulted in murder in the Amazon. Based on an analysis of growth rings, the same trees are providing scientists with a crucial record of climate change, one bolstered by the decline of coral reefs, the migrations of a diminutive species of butterfly, the shrinkage of glaciers, and ice cores laboriously collected at both poles. Moreover, scientists are using the most advanced computers to create models of climate change in the future, a story told in the closing stages of this work. In the chapters "Signs and Portents" and "Scenarios," we glimpse what the future may be like should global warming continue unchecked. Is the most recent incarnation of El Niño a preview of things to come—drought, fire, disease, torrential rains, mudslides, and oppressive heat? Or is global warming, as an outspoken minority of scientists steadfastly maintain, a good thing—a harbinger of more greenery, less privation, and freedom from cold? Finally, we visit Kyoto and examine the fierce debate surrounding the global warming treaty hammered out at the midnight hour, a treaty that the U.S. Senate may not ratify even though the United States leads the world in the production of greenhouse gases.

No one knows when the debate on global warming will be resolved by scientists, but it is one with which every citizen should be familiar. How we got from Jean-Baptiste-Joseph Fourier two centuries ago to our current predicament makes for a fascinating albeit deeply sobering tale—and, one hopes, for interesting reading as well.

Part One
THE TIME TRAVELERS

The past is but the beginning of a beginning, and all that is and has been is but the twilight of the dawn.

—H. G. Wells, *The Discovery of the Future*

THE GUILLOTINE AND THE BELL JAR

~

Where it is a duty to worship the sun it is pretty sure
to be a crime to examine the laws of heat.

—John Morley, *Voltaire*

Illness had been the old man's constant companion ever since his fourteenth year, when he spent his nights hidden in a large cupboard seeking to master the mathematics of Newton and Pascal by candlelight. Now, as the cycle of life played itself out, the gifted thinker who had originated the idea of global warming found himself back inside a wooden box, a device he used because he was so weakened by chronic rheumatism that to bend over was to risk a fatal attack of breathlessness. The human container had been built according to his own instructions. It kept his body upright by allowing only his head and arms to protrude, thus enabling him to work on his scientific papers to the last, even as he doggedly engaged in the voluminous correspondence required of the permanent secretary of the Académie des Sciences.

Visitors to his Paris apartment opposite the Luxembourg Gardens noticed something else. Their small and slightly built host kept the temperature extremely high, indeed almost tropical. It reminded one of Egypt, a place Jean-Baptiste-Joseph Fourier knew only too well. Ever since his return from that afflicted land in the wake of Napoleon's debacle more than thirty years before, his interior thermostat had never readjusted itself, suggesting that he might have been a victim of myxedema, which lowers the basal metabolic rate. Whatever his malady, it had been years since he had ventured forth without an overcoat and a servant bearing another in reserve, even in July and August.

As the end neared, Fourier refused to despair. He was too old to interest fate, which had long since exhausted the many unforeseen twists it had in store for him. He yielded instead to the embrace of nostalgia, allowing his thoughts to carry him back over the accumulated images of a lifetime.

Fourier had been born sixty-two years earlier, in 1768, in Auxerre, a Burgundian cathedral town situated on the heights overlooking the river Yonne. His father was a master tailor named Joseph, who produced at least fifteen children, the last dozen by Edmie Germaine LeBegue. Little is known of his mother, Edmie, except that she died at the age of forty-two, when Jean was only nine. After placing Jean and another child in the care of the local foundling hospital, Joseph followed his wife to the grave a year later.

Luckily for the orphaned Jean, the timely intercession of a townswoman saved him from a life of servitude. A Madame Moitton recommended the youth to the bishop of Auxerre, who in turn enrolled him in the local military school run by the Benedictines. An intellectual by nature if not by nurture, Jean blossomed under the tutelage of the monks. He was soon composing verse of exceptional quality for one so young, and he fell in love with mathematics, which became his passion. One night, a monk making his rounds spied a flame through the keyhole of a classroom cupboard. Fearing fire, he flung open the doors to discover the young Fourier solving equations by the light of discarded candle ends.

After rising to the head of his class and claiming his school's top prizes in rhetoric and mathematics, Fourier, who had already begun to suffer the symptoms of insomnia, dyspepsia, and asthma, decided to enter the Church. In 1787 he arrived at the Benedictine abbey of St. Benoit-sur-Loire to prepare for his vows while instructing the other novices in mathematics. Yet his mind was elsewhere. Having renamed himself Joseph after his father, he wrote of the desire to dream Descartes's dream—a reference to the supposedly mystical

Jean-Baptiste-Joseph Fourier
(Used by permission of The Granger Collection, New York)

formulation of analytical geometry. Torn by the conflicting demands of mind and spirit, Fourier added a wistful postscript to a letter addressed to one of his few scientific correspondents in late March of 1789: "Yesterday was my 21st birthday, at that age Newton and Pascal had acquired many claims to immortality."

Two weeks later, the Estates-General assembled at Versailles for the first time in 175 years. In the beginning Fourier appeared little interested. Not only was he still too young to speak in public, but he did not believe that the old tyranny would be replaced by anything better: "A new usurper tends to pluck the scepter from his predecessor," he wrote. Furthermore, he had reverted to his youthful habit of working into the small hours, and his health was suffering because of it.

In late October the newly established National Assembly issued a decree forbidding the taking of any further religious vows, the first step toward the abolition of the monastic orders. Fourier, who had not yet committed himself to the

Church, bid farewell to St. Benoit and returned to Auxerre to teach mathematics at his old military school.

Meanwhile, the Paris physician Dr. Joseph-Ignace Guillotin, a recently elected deputy to the National Assembly, had authored a novel proposal. Decapitation, Guillotin argued, should no longer be a right of the privileged classes alone, while the poor died by inches on the rack or the gibbet. Every person sentenced to death had a natural right to expire as swiftly and as painlessly as possible, "without torture."

This would be accomplished by the use of a machine whose origins might have dated from Roman times. Dr. Guillotin's modernized version was the very essence of simplicity: It consisted of two upright grooved posts surmounted by a crossbeam. Fitted into this wooden frame was a knifelike ax of forged steel. As the ax was released from a height, it gained momentum, slicing through the neck of the prone victim whose severed head conveniently dropped into a basket, a model of technological efficiency.

Guillotin cultivated the support of his fellow deputies, some of whom had already begun to address one another as "Citizen." Still, this was an enlightened age, calling for a certain amount of scientific experimentation. Several unclaimed corpses were beheaded in the hospital at Bicetre. Louisette, or Louison, as the machine was called, performed admirably and soon won the approval of the Assembly. On April 25, 1792, on the Place de Grève, the highwayman Nicolas Jacques Pelletier had the dubious distinction of becoming its first living victim. Dr. Guillotin, who by now had presciently retired from politics, was toasted throughout Paris, while his instrument of mercy was hailed by the habitués of executions as la guillotine.

Two short years later found Joseph Fourier in prison despairing of his life, all dreams of scientific immortality forgotten. His eloquence as a speaker and growing sympathy with the revolutionary cause had moved him to become involved in local politics. He had helped raise money, recruits, and horses for the war with Europe, then accepted the offer of a position on Auxerre's Committee of Surveillance, which, like dozens of others throughout France, had originally been entrusted with the relatively innocuous task of keeping an eye on passing strangers. Then came the Terror, forced on an inept government and National Convention by near famine conditions and the widening military threat. The powers of the local committees were expanded by the revolutionary tribunal in Paris as the nation was sucked into the vortex. One of Fourier's intellectual heroes, the mathematician and philosopher Condorcet, an original supporter of the revolution, had defied the radical Jacobins and died in prison. Another, Antoine Lavoisier, the genius and founder of modern chemistry, was guillotined. Fourier attempted to resign from the Committee of Surveillance, but his request was denied, giving rise to suspicions of the darkest kind in the minds of zealots who questioned his motives.

With one half of France in mourning for the other, Fourier was arrested for his vigorous public defense of three local families who had run afoul of the authorities. Anticipating his fate, he had traveled the 180 kilometers to Paris to plead his case before none other than Maximilien Robespierre, the guiding force behind the twelve-member Committee of Public Safety. Prim, chaste, and puritanical, the terror within the Terror listened politely as Fourier delivered his eloquent self-defense. A smile occasionally crossed the revolutionary's thin lips, for the onetime lawyer from Arras was partial to a facile tongue. Unfortunately, the fanatically idealistic Louis de Saint-Just, who also attended the proceedings, was not. A favorite of Robespierre and member of the committee, Saint-Just refused to be moved: "Yes, Fourier speaks well," he sneered, "but we no longer have any need of musical patriots."

Fourier was arrested on July 4, 1794, released, then rearrested a few days later as the so-called Great Terror reached its climax. The insatiable guillotine was devouring its daily offering of victims with an efficiency that exceeded every prediction, including that of the public prosecutor Antoine Fouquier-Tinville, who spoke proudly of heads falling in assembly-line fashion, "like tiles."

It was Thermidor (Heat), the eleventh month of the French revolutionary calendar, and Fourier could expect the end at any moment. Then, amazingly, word arrived by courier from Paris that Robespierre, Saint-Just, and their fellow Jacobin Georges Auguste Couthon had been shouted down by the National Convention and arrested. The next morning, after witnessing a screaming Couthon struggle mightily for fifteen minutes and Saint-Just submit without a word, a semiconscious Robespierre, who had been wounded by a bullet in the face during his arrest, was dragged up to the platform and strapped onto a plank. The bandage was torn from his shattered jaw before his head was pushed through a little window, causing him to cry out in pain seconds before the steel blade descended. Robespierre had fallen exactly one year to the day after joining the Committee of Public Safety.

The deserted convents, abandoned chateaux, vacated schools and warehouses that served as makeshift prisons were soon emptied of their political captives. A liberated Fourier offered up a silent prayer of gratitude and made plans to depart Auxerre as soon as the opportunity presented itself. There was no trusting the people among whom he had been born and grown up, nor for that matter anyone else.

If ever Fourier was to establish a permanent place for himself in science, or natural philosophy as it was then known, the road lay through Paris, the mecca for all aspiring young Frenchmen. While the revolution had claimed certain of his heroes, Pierre-Simon Laplace and Joseph-Louis Lagrange were still very much alive, plumbing the firmament with telescopes, calculating the movements of the planets, comets, and tides as they dispelled any lingering doubts that Newton's law of gravitation was superior to the vortices of Descartes. At the same time, Fourier hoped to shed his revolutionary past by donning the cloak of anonymity.

In the spring of 1795, he wrangled a nomination to the École Normale, recently created by the National Convention for the purpose of redressing the nation's acute shortage of elementary school teachers. There he finally set eyes on Lagrange, "the first," he wrote, "among European men of science." The mathematician Laplace, who was also among the first rank of natural philosophers, was less appealing, perhaps because of the his obsequious manner in the presence of authority.

Just as Fourier was getting his feet on the ground, rumors began circulating of actions against him back in Auxerre for the part he had played in the revolutionary government. On the night of June 7, 1795, he was awakened by armed guards and marched off to prison in the Rue des Orties, having scarcely been given time to dress. As he was being led away, the concierge expressed the hope that Fourier would return soon, prompting the chief of the armed guard to retort, "Come and get him yourself—in two pieces!"

As in Auxerre, Fourier was released within a short time of his imprisonment, then rearrested weeks later in the middle of the night. This time the authorities charged him with terrorism, of being a Jacobin and abettor of the same Robespierre who had denied Fourier's earlier appeal on the grounds that a lukewarm revolutionary was a luxury France could ill afford.

What happened at this point is a matter of conjecture. By now Fourier's gifts as a mathematician had become known to Lagrange, Laplace, and Gaspard Monge, the leading geometrician of the day. Indeed, Fourier was enrolled in Monge's course in descriptive geometry at the newly opened École Polytechnique, whose standards were more rigorous than those of the École Normale. It seems likely that Monge, perhaps with the support of others, gained Fourier's release, aided by a changing political climate. No sooner was he back out on the street than the Convention was attacked by a proroyalist mob, in October of 1795. An obscure corporal named Napoleon Bonaparte, who had himself been briefly imprisoned during the Thermidorian reaction, scattered the assailants with what he contemptuously dismissed as "a whiff of grapeshot."

In March of 1798, Citizen Fourier, whose reputation as a mathematician had grown during his three years in Paris, gaining him a professorship at the École Polytechnique, received a letter from the minister of the interior. The Republic was in need of his "talents" and "zeal" for an unspecified mission in a foreign land. One month later Fourier stepped onto the deck of a ship for the first time in his life, joining the company of Napoleon, his fellow generals and officers, some 200 scientists known as the *corps des savants* led by Monge, and 50,000 soldiers and sailors, all crowded onto 180 vessels destined for they knew not where.

The French armada passed through Gibraltar with the English in hot pursuit. Horatio Nelson's squadron of thirteen seventy-four-pounders came as close as two miles to them on

June 22, but their luck held and they reached Malta without incident. The island's long-resident Knights of St. John were easily subdued and forced into exile, enabling Napoleon to add 7 million gold francs to his burgeoning war chest. On July 1, 1798, the armada sighted Pompey's pillar at Alexandria, and the city was taken the following day after little more than token resistance.

It was in the land of the Sun-worshiping pharaohs that Fourier's scientific interest in heat was kindled. Yet he was too preoccupied by his duties as secretary of the newly formed Institute d'Egypt to carry out much research on the subject.

After his return to France in 1801, he wanted nothing more than to resume his work at the École Polytechnique, but Napoleon had different ideas. The general had spotted Fourier's administrative genius and appointed him prefect of the department of Isère, with headquarters in Grenoble. There Fourier would remain for the next twelve years, so pleasing his diminutive benefactor that Napoleon made the onetime egalitarian a baron.

Meanwhile, Fourier had tackled the complicated question of how heat spreads, or what physicists call diffusion. Working with objects of various shapes, he developed the now famous "diffusion equation," which expresses the movement of heat within the body itself. No matter what form he was working with—be it a cylinder, sphere, rectangle, or ring—he was able to account for the phenomenon mathematically. To his lasting credit, Fourier then carried his work a step farther. He developed a second equation to deal with heat movement on the surface of an object, the so-called boundary value.

In December of 1807, a hopeful Fourier presented the fruits of his concentrated labors to the Académie des Sciences in the form of a long paper. Of the four examiners, Monge, Laplace, and Lacroix opted for its publication. However, Lagrange voiced strong objections to it on the grounds that the mathematics, now known as the Fourier series, contradicted Lagrange's own trigonometric formulations. Unfortunately, one outweighed three in the eyes of the Académie, so the paper went unpublished.

Fourier got a second chance when, in 1810, the Académie offered a prize to the person who could successfully solve a problem on heat diffusion. He revised his earlier paper, adding a new mathematical analysis on heat flow in infinite bodies, and sent it off to Paris. The prefect's entry, titled "The Analytical Theory of Heat," won the competition, but not cleanly as it turned out. Lagrange launched new criticisms on the grounds of "rigor and generality." Fourier reacted by penning a withering reply via Laplace. The Prize Essay, as it became known, was not published, nor would it be until 1822, well after Lagrange was in his grave. Yet history was on Fourier's side. The powerful mathematical tools he invented were destined to motivate much of the leading work in that field for the remainder of the century and beyond, spilling over into physics, theoretical astronomy, and engineering.

With his niche in the pantheon of scientists secure, Fourier began thinking on a grander scale—the laboratory of the macrocosm. In Egypt he had watched Bedouin traders, their lumbering camels in tow, disappear into the Sun's blinding golden eye while their wavering images were reduced to vapor. Constant, eternal, without mercy, the giant fireball heated the sands until, by midday, the desert surface cut and burned like the blade of a newly forged knife. Long before Napoleon's troops were frozen stiff in the Russian winter, fully half of his expeditionary force baked to death under the African Sun, "a great sacrifice," someone remarked, "for a pretty rock," meaning the Rosetta stone.

One day in the early 1820s, Fourier began to ponder the question of how Earth stays warm enough to support the diverse range of flora and fauna inhabiting its surface. Why is the heat generated by the Sun's rays not lost after striking and bouncing off the great oceans and landmasses of the world? Taking pen in hand, he set down a novel hypothesis. Much of the heat does in fact escape back into the void, but not all. The invisible dome that is the atmosphere absorbs some of the Sun's warmth and reradiates it downward to Earth's surface. Fourier likened this thermal envelope to a domed container made of glass, a gigantic bell jar formed out of clouds and invisible gases. In coming together, the water vapor and other gases simulate a vault that receives and conserves heat, without which all life would surely perish.

Fourier shaped his thoughts into the article "General Re-

marks on the Temperature of the Terrestrial Globe and Planetary Spaces," which the editors of the *Annales de chimie et de
physique* were only too happy to publish in 1824. Three years
later, virtually the same article appeared in the pages of *Memoires de l'Académie Royale des Sciences.* Because of its speculative nature, the bell jar hypothesis was not considered
Fourier's best work or his most memorable.

Only with the onset of the industrial revolution would
this French genius's obscure papers come to light. And even
then the phenomenon of global warming was thought of in
the most benevolent of terms. What would it matter if gases
emitted by Europe's giant smokestacks raised the temperature
of Earth's atmosphere by a few degrees, a process that would
almost certainly take many centuries to unfold?

Thus we are left with the surreal image of an old man,
draped in thick woolen clothes and heavy blankets, propped
up in his boxlike chair, slaving away to the end, his hand so
cramped as to render his script unreadable. The honors he
had dreamed of when young were now his without the asking:
membership in the Royal Society of London, elevation to the
Académie Française, the presidency of the council of the École
Polytechnique, and more.

As death approached, he wrote a friend of having already
glimpsed "the other bank where one is healed of life." Just as
he wished, the call came suddenly on May 16, 1830, when he
suffered a fatal heart attack late in the afternoon. The funeral
took place two days later, with burial in the cemetery of St.
Père Lachaise on the outskirts of Paris. Fourier had chosen the
spot himself; it was close to his revered master Gaspard
Monge, the companion of his long days in Egypt, where he
had always been warm.

2

THE CRYPTIC MOTH
~

Until recently the great majority of naturalists
believed that the species were immutable
productions, and had been separately created.

—Charles Darwin, *The Origin of Species*

England's amateur naturalists of the nineteenth century
formed a large and dedicated fraternity. Still, not one of them
so much as suspected that a recently documented change in
the color of an insect was a sign that the world, together with
its climate, was being forever altered. Among the fraternity's
more avid members was the lepidopterist R. S. Edleston,
whose well-documented observations on butterflies and
moths occasionally made their way into the pages of such scientific journals as *Entomologist* and *Zoologist*. Yet each time
the bespectacled Edleston peered into the glass-covered case
on his wall, he was haunted by the same recurring question:
What strange alchemy had given rise to the pinioned and
mute creature before him?

Edleston lived about a mile from the center of Manchester, whose population had risen from some 10,000 early in the
eighteenth century to 300,000 by 1848, when the collector's
story began to unfold. He had already gathered and carefully
preserved many of England's estimated 780 species of larger
moths, and he knew their major characteristics well. In contrast to butterflies, which are active during the daylight hours,
moths emerge at twilight and retire before sunrise, a trait

known to zoologists as crepuscularism. Moths have thicker bodies and duller wings than their more slender and brightly colored cousins, while their prominent antennae, which appear featherlike under a magnifying glass, are rarely knobby at the tips, like those of the butterfly. Below the antennae are a pair of large compound eyes composed of many light-sensitive elements, each with its own refractive system and each forming a portion of an image. Among the most intriguing aspects of the moth's anatomy is the unusual mouth concealed by a dense array of scales. The proboscis is coiled like the mainspring of a pocket watch when not in use. When extended, it becomes a hose in reverse, siphoning liquid food directly into the pharynx of the digestive system.

Among the many species represented in Edleston's collection was *Biston betularia*, commonly known as the "peppered" moth because its light gray wings and body are covered with irregular black spots, part of the natural camouflage essential to its survival. This protective mechanism classifies it as a cryptic species, one that has developed an extraordinary resemblance to certain background objects in its environment. Thus Edleston was stunned when he captured a specimen of *B. betularia* unlike any he had ever seen. In place of the familiar gray-and-black markings, the collector was gazing on a moth possessed of wings so uniformly dark that he termed it "negro."

Here matters would have stood, but for one thing. In the years that followed Edleston's discovery, other naturalists were initiated into the brotherhood of the cryptic moth, and their reports were duly recorded in the section of the *Entomologist* devoted to readers' queries. In 1864 Edleston made the startling announcement that the black aberration of *B. betularia*, almost unknown sixteen years earlier, now made up the majority of wild males sighted near Manchester. "If this goes on for a few years," he predicted, "the original type . . . will be extinct in this locality." Equally baffling to collectors was the fact that similarly marked specimens had begun to appear among other species of moths, as well as ladybird beetles and spiders. In recognition of this development, the dark form of *B. betularia* was rechristened *carbonaria*.

~

Far to the southeast of industrial Manchester, in the bucolic hamlet of Down, a pain-racked Charles Darwin was completing the sequel to his 1859 magnum opus *On the Origin of Species by Means of Natural Selection*. The *Origin* had been preceded, in 1842, by a 35-page sketch of Darwin's evolutionary theory, which he expanded into a 230-page essay two years later, a painful emotional process he once likened to "confessing a murder." The master of Down House was a gravely different man from the generally healthy, exuberant naturalist of the HMS *Beagle*. He suffered from dizziness, headaches, eczema, fatigue, and what he termed "many bad attacks of sickness" that caused him to be confined, like Fourier, to a tattered blue chair especially designed for semi-invalids. Fearing that death would overtake him before he could publish, he had drafted a letter to his wife, Emma, asking her to see his work through the press in the event of his early demise.

All of that suddenly changed when, in July 1858, Darwin received a short but stunning manuscript from the naturalist Alfred Russel Wallace, sent from Malaysia. It was a comparative study of the biology of Brazil and the East Indies based on the same evolutionary theory Darwin had developed some

Biston betularia *(top) and* **Biston carbonaria**
(From *The Evolution of Melanism* by Barnard Kettlewell,
published by Clarendon Press)

twenty years earlier. He cursed himself for having waited so long to publish, then did the gentlemanly thing by seeing to it that Wallace's paper was printed simultaneously with excerpts from his own 1844 essay. The *Origin* followed a year later.

Darwin was nothing if not exhaustive. It had long been his dream to publish a second book that would account for the gaps in the intricate chain of evidence supporting evolution by natural selection—and to read humankind into the record. His deep-set eyes, dark and haunted after decades of intensive labor, reveal the outcome of that unfulfilled quest. The intellectual revolutionary had no direct knowledge of the mechanism of inheritance, nothing adequate to account for either the appearance or the preservation of favorable variations from one generation to the next. After much agonizing, he finally settled on a theory known to his fellow naturalists as pangenesis. According to its tenets, each tissue, perhaps each cell within the adult organism, produces minute particles called "gemmules," which pass through the body and concentrate in the reproductive organs. When fertilization occurs, the gemmules from both parents blend, thus determining the specific character of the newly formed embryo.

Sadly for Darwin, there was the bad to contend with as well as the good. After the faith-shattering loss of his beloved daughter Annie, he was haunted by the thought that his own chronic illness might one day crop up in his other offspring: "My dread is hereditary ill-health," he wrote his clergyman cousin William Darwin Fox. "Even death is better for them."

Darwin's reproductive theory was seriously called into question by, of all people, a Scottish engineer named Fleeming Jenkin. In a belated review of the *Origin*, Jenkin contended that the process of blending would effectively swamp variants favorable to survival. For example, if an individual born with a significant new characteristic, or "saltation" in the language of the time, breeds with a normal partner, their offspring will inherit only one half of the variant. The next generation will possess only one quarter, and so forth until the dilution is complete.

Loath to discard the pangenesis theory, Darwin cleverly sidestepped Jenkin's critique by revising his model of variation. He postulated that, in contrast to a single individual, sev-

Charles Darwin
(Used by permission of Culver Pictures)

eral breeding pairs might have the same saltation. If, indeed, this is the case, then swamping would not occur even if inheritance does involve blending. Still, the theory left him with grave self-doubts. His detractors would surely class it as another of his "mad dreams, wildly abominably speculative."

Even more vexing to Darwin was the fact that, try as he might, he was unable to point to a single living example of evolution at work in nature, all the intriguing fossil evidence in the world notwithstanding. The case he had made for evolutionary gradualism was so persuasive it seemed impossible that a new variant, however advantageous, could spread through a species within the lifetime of a single observer like himself.

The only circumstances under which this had happened was in the breeding of such domesticated species as the pigeon. In the mid-1850s, Darwin himself entered the game by having a wooden loft built next to rambling Down House. He installed breeding pairs of fantails, with their great erectile feathers; gravity-defying tumblers, which turn backward somersaults in flight; pouters, with enormous crops; feisty carriers and laughers who, Darwin quipped, "wouldn't laugh." To further his research, the stay-at-home country gentleman traveled to London, where he was befriended by grubby vendors, deformed weavers, and weathered ferrymen who could "pinch, bustle, and crown birds," literally creating forms as stylish and as subject to change as the latest female fashions. The problem was this: "Selective transmutation," as Darwin called it, had resulted from human manipulation. What could be accomplished by pigeon fanciers using artificial means was not directly applicable—at least to the same degree—to nature, where gradualism reigned supreme. Yet he also realized that by selecting the "best" individuals, breeders were performing a function not unlike that occurring in the wild.

Darwin was in his early forties when the first black peppered moth was captured in 1848, and he was still at the height of his powers when Edleston announced that black males now outnumbered their gray counterparts around Manchester. Yet there is nothing in Darwin's notes to suggest that he ever read of *carbonaria*'s dramatic appearance and proliferation. The black moth was soon to become part of what the twentieth-century entomologist Bernard Kittlewell termed "Darwin's missing evidence."

The Descent of Man and Selection in Relation to Sex went to press in 1871 and was marked by the same flaws as the *Origin*, much to its author's regret. Still, there was no denying the power of Darwin's central argument that the survival of lifeforms favors those that are best adapted to their environment, while the less suited ones die out. When he passed away eleven years later, at age seventy-three, the man who had been called every derogatory name imaginable, including Ratcatcher by his exasperated father, was interred in Westminster Abbey with due pomp and circumstance, a towering icon of the waxing Victorian age.

~

Revolutions are made in large places—the whole of France or Russia—and in small ones, such as a naturalist's study or a monastery garden tucked neatly behind cloistering walls. In a room overlooking his carefully tended peas, which he sometimes referred to as his children, an Augustinian monk made extensive notes in the margins of his German translation of Darwin's *Origin*. The concept of evolution by natural selection made a good deal of sense to Father Gregor Mendel, who possessed a kind if wistful face. Yet when it came to the matter of heredity, he was certain that Darwin had gotten it wrong.

The experiments that would one day lead to the science of genetics were begun at Brünn, Moravia, in 1856. Mendel carefully surveyed the available varieties of peas before settling on the genus *Pisum*, whose distinct characteristics such as seed shape, seed color, height, and flower location could be tested by hybridization, or crossbreeding. Eight years of tedious experiments followed, during which Mendel grew and analyzed at least 28,000 individual plants, whose seeds he gathered, counted, sorted, and labeled, sometimes assisted by fellow members of the local Natural Science Society who liked to stop by and talk shop.

It now appears that Mendel rejected Darwin's model of pangenesis, or blended inheritance, from the beginning. Instead, he theorized that each parent supplies one of two hereditary particles he termed *elements*, which form pairs. Using careful techniques of artificial fertilization, he demonstrated that in the first hybrid (F1) generation, one character out of each pair—tallness, for example—appears in all the offspring, causing them to look the same. The other character—in this instance shortness—has seemingly disappeared. But when Mendel self-fertilized these hybrids to produce a second (F2) generation, something remarkable happened. Three of the resulting plants were tall as before, but the fourth was short. Duly impressed, he decided to call the prevalent characteristic *dominant*; the other, which remained latent and appeared in the second generation, he termed *recessive*. So it was with other generations of hybrids, where the famous 3:1 ratio held. He had shown that the distinct elements from each parent are transmitted to their offspring without change, like the hardened peas gathered from the very plants he had so painstakingly tended. The celibate monk had fathered the first scientific law of living forms.

"To this day," wrote the scholar Robert C. Olby, "an air of mystery surrounds the name of Gregor Mendel." This doubtless has much to do with the cleric's quiet frustration at not gaining a scientific audience for his revolutionary labors. His now famous paper "Experiments in Plant Hybridization" was read in two parts before the Natural History Society of Brünn, in February and March of 1865, where it attracted little attention for three reasons: The society was composed mainly of amateurs; Mendel was not a famous scholar but a physics teacher in the local high school; and few were conversant with certain obscure mathematical equations he employed in his experiments.

The following year, the forty-eight-page paper that was Mendel's magnum opus appeared without alteration in the *Proceedings* of the Natural History Society. Some 140 copies were mailed to scientific institutions in various countries, and the author commissioned another 40 reprints for distribution as he saw fit. Darwin never laid eyes on the article that would have served as the consummation and confirmation of his life's work, providing he had fully grasped its profound implications, which was by no means assured. No other scholar of Mendel's day appears to have understood his spare mathematical logic or appreciated the fact that knowledge garnered from lowly hybrid garden peas had forever altered humankind's understanding of the natural world. Not until 1900, sixteen years after Mendel's death, was his paper rediscovered by a trio of scientists working independently of one another. Even so, it would take thirty years beyond that before the theories of natural and genetic transformation were fully synthesized. Yet Mendel, no less than Darwin, knew what he knew. As the disappointed monk lay dying from uremia and dropsy, he is said to have exclaimed to the nun attending him, "My time will come!"

~

When seen from a distance, Manchester took on the appearance of a giant cloud commonly mistaken by the approaching traveler as a harbinger of rain. On entering the city, the stranger was greeted by darkness at noon. Surrounding the sprawling, many-windowed factories were hopeless streets of tiny row houses, whose occupants, lamented Elizabeth Gaskell, a resident novelist, "only wanted a Dante to record their sufferings."

With the denuding of England's once great forests, coal

Father Gregor Mendel
(Used by permission of The Granger Collection, New York)

fueled the factory boilers and warmed the hearths of the workers when they could afford it. The effects of the atmospheric pollution on human health had not gone unnoticed. As early as 1844 legislation had been passed to control the smoke emitted by the city's industrial furnaces, whose chimneys, in the words of one awed observer, "rose ever higher into the firmament, like the arms of God."

To no one's surprise, the laws went largely unheeded. Indeed, the illegal carboniferous emissions were wryly renamed "unparliamentary smoke." Nor did the law do anything to curb the burning of coal in private residences. By 1866 the city's medical officer pronounced Manchester's citizens among the most unhealthy in Britain and attributed their condition to the foul air they breathed.

Yet within this blighted realm were weavers who worked their looms with open copies of Newton's *Principia* and Linnaeus's *Systema naturae* propped against the wood frames. Mathematical problems were studied with interest by broadspoken and common-looking young men. There were self-taught botanists who knew the name and habitat of every plant within a day's walk of the city, weekend entomologists who carried homemade nets over their shoulders, ready to snag any insect that chanced to cross their path. Surely the patrician befriended by London's lowly pigeon breeders would have fared equally well in the company of such men. For near Manchester evolution was occurring not in the thousands of years Darwin described in the *Origin* but in thousands of days—not in hundreds of generations but in mere dozens.

Each year as many as fifty tons of industrial fallout settled over every square mile of the city and the surrounding countryside. The smoke particles coated the leaves and twigs of the trees and shrubs, impairing their function as natural filters. When summer storms came sweeping over the Pennines, the pollutants cascaded down the branches and bows, creating "rain runs" that destroyed the scaly lichens attached to the bark—lichens whose coloration could hardly be distinguished from the mottled wings of the peppered moth. Nothing grew at the base of the dark and naked trunks, where the top few inches of surrounding "earth" consisted of little more than acrid soot. Even the naturally gray rocks protruding above the surface had turned a permanent black.

A naturalist with the Dickensian name J. W. Tutt is credited with making the connection between the first great man-made change in the environment and the alteration of a species. Tutt, who published widely on British moths in the late nineteenth century, believed that the common form of *B. betularia* had virtually disappeared due to the alteration of its habitat.

This moth, like most related species, flies at night and passes its days motionless on the branches of rough-barked deciduous trees, such as the oak. Lichen-covered rocks and decomposing timber also provide safe harbors. So long as the environment remains stable, the survival rate is sufficient to sustain the species, whose larvae feed on foliage from June to October before entering the pupal state after burrowing into the ground. But when gray branches turn black and lichens disappear, the once cryptic moth is suddenly exposed and ruthlessly eliminated by predators, mostly of the feathered variety. This is a classic example of survival of the fittest. Further support for Tutt's theory was later supplied by a study of England's rapidly changing industrial geography. *B. betularia*'s demise followed directly in the wake of the polluted winds that crossed the country from the northwest to the southeast.

Enter *carbonaria*, possessed of a gene, or in Mendel's words "dominant element," that alters the dark pigment (called melanin) in its wings. While the other members of its species were rapidly becoming the victims of pollution, *carbonaria*'s somber camouflage was well adapted to the changing environment. The insect was soon outproducing its light-colored brothers to the point where, by 1900, it constituted some 90 percent of the peppered moths living near Manchester, whose human population now numbered over half a million. As scientists were soon to learn, this phenomenon was not confined to the British Isles. Industrial melanism was about to surface in the valley of the Rhine and in the Netherlands, then eastward near Hamburg in northern Europe and southward into what became Czechoslovakia and Poland. And though somewhat delayed, it was only a matter of time until the melanic form of the peppered moth was observed in North America, not far from where Henry Ford, the apostle of mass production, had set up shop. Contrary to Darwin's belief, a species in the wild could be drastically altered at the hands of humans.

3

"ENDLESS AND AS NOTHING"

~

Time has no divisions to mark its passage, there is
never a thunderstorm or blare of trumpets to announce
the beginning of a new month or year. Even when a
new century begins it is only we mortals who ring
bells and fire off pistols.

—Thomas Mann, *The Magic Mountain*

If plants and birds and insects could be altered—whether intentionally or not—by human tinkering, then what of Earth
itself? The answer to that question very much depended on
the resolution of another: Just how old is the planet that most
believed had been entrusted by its creator to man?

By the time Darwin died in 1882, scientists knew the
world to be an ancient place with primordial roots, but it had
not always been thought so. More than two centuries earlier
James Ussher, the exiled archbishop of Armagh and one of the
leading Christian scholars of his time, established a much
briefer timeline. By totaling up the ages of the post-Adamite
generations as set forth in the Bible, Ussher determined that
Earth was created around noon on October 23, 4004 B.C.
Moreover, Judgment Day, which would bring earthly time to
an abrupt end, was thought by the archbishop to be close at
hand, for signs and portents abounded.

The calculations of Ussher and those of a similar mind
gave rise in geologic circles to the theory of creation known as
catastrophism. At intervals in Earth's history, it was argued, all
living things had been destroyed by great upheavals ordained
by God—floods, earthquakes, volcanoes—each of which had

worked their will on the planet's surface, creating mountains, valleys, and vast seas. Humans, together with all other living species, appeared only after the latest of these disasters, the great Noachian Deluge of some 6,000 years ago. Fossil evidence of vanished life-forms was simply the residue of creations and calamities from the prehistoric unknown, before biblical time. Among catastrophism's most ardent champions was Fourier's friend and colleague the baron Georges Cuvier. As the most revered geologist and comparative anatomist of his day, the Frenchman easily overrode all opposition to the theory until well into the nineteenth century.

One of the earliest of the lonely dissenters was James Hutton, a retiring Scotsman with the spare and elongated countenance of a medieval ascetic. Hutton's merchant father had died when his son was quite young, leaving James to his own devices. The youth developed an early interest in chemistry but was forced to enter the legal profession as a means of earning a living. Rather than apply himself to the law, Hutton is said to have spent most of his time entertaining his fellow apprentices with scientific experiments, conduct for which he was soon dismissed.

Apparently unfazed, he joined forces with his friend James Davie, and together they developed a chemical method of extracting sal ammoniac (ammonium chloride) from coal soot, which was increasingly abundant in the 1750s. The manufacture of the compound for use in the burgeoning textile industry and in certain medical treatments provided Hutton with the financial wherewithal to follow his bliss.

Hutton was never more content than when hiking over some nondescript hill or plumbing the darkest reaches of a coal mine. He collected specimens of rocks and minerals from across the British Isles and extended his research into the countryside of northern France, Belgium, and Holland. Lacking company, he delivered soliloquies to the surrounding stones, which answered him by revealing their secrets. Hutton became such an authority that later generations would declare him the founder of modern geology.

Believing that the facts of Earth's history were to be found in natural processes rather than in the human record, Hutton considered the Bible an unreliable tool for determining the age of creation. After examining the structure of many types of rocks, he advanced the principle that pressure and extreme heat beneath the planetary surface are the keys to understanding geologic time. As land surfaces wear down under the constant assault of wind and weather, the eroded sediment is borne by rivers to the world's oceans, where it settles into the depths. Over time it is recycled by volcanic activity, giving rise to new rocks and mountains in a ceaseless pattern scarcely fathomable in human terms. "Time," a prescient Hutton wrote in his multivolume *Theory of the Earth*, "is to nature endless and as nothing."

In place of a divinely created world measured in human generations, Hutton formulated the concept of uniformitarianism, which holds that nature is a perpetual and circulatory worker independent of the will of God. The processes of erosion, fusion, and volcanism that scientists chronicle today are no different from those that unfolded in eons past. Nature's history is not one of ordained catastrophes but of continuous decomposition and regeneration, time out of mind. Just as the astronomers of the scientific revolution had shattered the inner dome of heaven, raising the prospect of infinite space, James Hutton saw boundless time in the dust of the air and the clay that stuck to the soles of his sturdy walking shoes. "The way," as one historian of science observed, "was opened for Darwin."

~

During his student days at Cambridge, Darwin had developed a reputation as "the man who walks with Henslow," a reference to John Stephens Henslow, professor of mineralogy and botany. As Darwin's teacher and friend, Henslow was responsible for recommending the young man as naturalist and gentleman's companion to Captain Robert Fitzroy, commander of the HMS *Beagle*. For a parting gift, Henslow gave Darwin a copy of Alexander von Humboldt's *Personal Narrative*, an engaging account of the renowned explorer's expedition to Central and South America, for which Darwin himself was bound. Henslow also recommended that Darwin pick up the first volume of Charles Lyell's recently published *Principles of Geology*, though he cautioned him not to believe everything Lyell had written. Weeks later, as the *Beagle* tacked southward in a hot,

James Hutton   *Charles Lyell*
(Both used by permission of The Granger Collection, New York)

seasickening breeze, the naturalist lay quietly in his cabin reading the *Principles* between seizures of nausea that sent him scurrying to the rail, much to the crew's amusement.

Lyell, who was born in 1797, the year Hutton died, was an equally staunch uniformitarian. His reading of *Theory of the Earth*, coupled with his own prodigious geologic research, convinced him that the planet has existed for countless millions of years rather than the paltry millennia of Christian time. But what set Lyell apart from Hutton were the observations contained in the second volume of his *Principles*, which was not yet out when Darwin left England and did not catch up with him until mid-1832.

Here Lyell considered the changes that have occurred in the living world across geologic time. In this he was aided by the construction and excavation spurred by the industrial revolution. The digging of canals and the opening of new mines provided the stratigraphic detail on which uniformitarianism was based. Geologists and engineers like George Scrope and the intrepid William "Strata" Smith inventoried hundreds of mineral layers, which Smith likened to "the ordinary appearance of superposed slices of bread and butter."

By focusing on the fossilized shells within such layers, Lyell was able to show how, as one moves from older to younger strata, one extinct species is succeeded by another until, near the surface, contemporary life-forms appear. To account for extinction, Lyell argued that the survival of each species is dependent on the continuance of certain physical conditions in its environment. Geologic processes tend to alter these conditions over time, both locally and across wide regions. A species may respond by shifting habitats, while in more extreme cases it may be obliterated together with others dependent on its existence. Thus the continuity of the fossil record is paralleled by the continuity of geologic change, a lesson that quickly hit home with Darwin: "Certainly," he wrote in *The Voyage of the Beagle,* "no fact in the long history of the world is so startling as the wide and repeated extermination of its inhabitants." Hutton had whispered into Lyell's ear and Lyell into Darwin's.

It was also Lyell who first named certain of the major geologic eras in Earth's long history, beginning with the Precambrian nearly 4,000 million years ago and reaching forward to the current Cenozoic, now some 65 million years in duration. Of these eras, none is more important to our story than the Paleozoic, whose several divisions include the Carboniferous period stretching back 300 million years.

Flourishing in the warmth and damp of vast coal-forming swamps were giant ferns and fernlike trees whose names evoke the lyricism of the muses: Lepidodendron and Sigillaria, slender trees one hundred feet tall with narrow, sharply pointed leaves thirty inches long; Cordaite, a towering ancestor of the modern pines and spruces; Calamites, ancient relatives of the common horsetail that stood five times the height of a man; and in the background, like a mute chorus, the less conspicuous but no less essential ferns—Pecopteria, Alethopteris, Linopteris, Ondantopteris, Neuropteris, and Mariopteris.

Although the Paleozoic is known as the Age of Plants, animals and insects too made their contribution to the giant compost heap covering much of the planet's surface: early amphibians, reptiles, spiders, millipedes, snails, and scorpions; enormous dragonflies with a wingspread up to twenty-nine inches, and—nightmare of nightmares—more than 800 species of cockroaches. From the shallow inland seas came fishes, clams, and many varieties of crustaceans; from the oceans mollusks, crinolids, urchins, and lime-making foraminifers. Within this murky, gurgling quarter-billion-year-old brew festered the origins of the industrial age.

Normally, when a plant or animal dies, it is attacked by bacteria and other microorganisms, causing it to decompose in relatively short order. But when it falls into a swamp, the result can be very different. Decay is curtailed by a lack of oxygen, and the partially rotted material becomes a sodden but coherent mass of combustible muck called peat. (Peat is still being formed in Canada, Russia, Scotland, and Ireland, where centuries ago young men with red hair and the beards of Druid princes were bled to death by a sacrificing priest, then cast into bogs whose chemical properties preserved every physical detail down to the smile lines gracing the corners of their eyes.)

In times of maximum coal formation, the great peat swamps were located on the seaward margins of the continents. These areas of low elevation were subjected to periodic flooding, during which successive layers of mud and sand were deposited over the peat. As each swamp was inundated, another formed atop the overlying sediment, creating layers that would eventually become consolidated by heat and pressure into coal seams ranging in thickness from a few hundred feet to upwards of two miles. Once the coal was formed, the sequential layers of coal-bearing rock were uplifted by spasmodic movements of Earth's crust that resulted from changes in the relative positions of the land and the sea. In the process the seams were warped into upfolds, called anticlines, and downfolds, known as synclines. With the passage of time, the upfolds were mostly eroded away by the elements, while the buried downfolds were preserved at depths ranging from several hundred to many thousands of feet. To the paleobotanist and paleontologist, heirs of Hutton and Lyell, has fallen the task of determining the age of these beds by studying the plant and animal remains locked within.

~

Anyone who has ventured into an introductory chemistry course knows that carbon is present in more compounds than all other elements combined. It exists in the most massive stars and is the source of their energy through a series of thermonuclear reactions; in archaeology and geology radioactive carbon-14 is used to determine the age of objects, both natural and man-made; and carbon's linkage in the 1940s with silicon has led to the development of a vast number of new substances, including the revolutionary semiconductors of the computer age. The element accounts for 18 percent of the human body by weight and is an indispensable constituent of all life as we know it, circulating as it does back and forth between the living and the nonliving in what biologists term the "carbon cycle."

As the central element in most compounds of which organisms are composed, carbon is derived from carbon dioxide (CO_2), also known as carbonic acid, a colorless, odorless, and tasteless gas found in the air. The process of incorporating carbon's inorganic molecules into the more complex molecules of living matter is called fixation and is accomplished by photosynthesis. Green plants combine CO_2 and water to form simple sugars, or carbohydrates, using the Sun's energy to trigger and sustain the chemical reactions involved. Some of the resulting carbon is incorporated in the plant tissue itself, while the rest is returned to the atmosphere or water by respiration. When the plant dies, its tissues are consumed by bacteria in the process called decay. By the respiration of these microorganisms yet more carbon is released into the atmosphere.

Unlike plants, animals are incapable of making their own carbohydrates. Thus grazing herbivores must rearrange and degrade the carbon compounds as part of their metabolic process. Some of these are stored in the animal's tissue, where they are passed on via the food chain to predatory carnivores for building their own cells. Respiration returns even more of the carbon to the atmosphere, and the cycle is completed when the animal, whether predator or prey, dies and its carbon compounds are broken down by decomposition and released as CO_2 for use by plants once more.

On a global scale the carbon cycle involves an exchange of CO_2 between two great reservoirs: the atmosphere, which contains about 1 percent of the total carbon pool; and the waters of the planet, in which are locked some 71 percent of the world's carbon. Atmospheric CO_2 enters Earth's oceans and lakes by diffusion across the air-water surface. If the concentration of CO_2 in the water is less than that in the atmosphere, it is diffused into water; if, on the other hand, the CO_2 concentration is greater in water than in the atmosphere, the CO_2 enters the air. Additional exchanges take place within aquatic ecosystems, where excess carbon combines with water to form carbonates that are deposited in bottom sediments. Some of the remaining carbon (4 percent of the planet's total) is incorporated into forest vegetation, where it may remain out of circulation for hundreds of years or even longer. Sooner or later, however, the slow smokeless burning of decay results in the accumulation of peat and, ultimately, of great stores of fossil fuels—coal, oil, and natural gas—that originally formed in the Carboniferous period.

The sedimentary record suggests that for roughly a half billion years between the end of Precambrian time and the beginning of the Cenozoic era, Earth possessed an unusually warm and moist climate. Temperatures at the equator were at least as warm as they are today, while the seas were open and the polar caps free of ice. Deserts existed on the continents, most likely in subtropical latitudes, and at times may have been more extensive than the Sahara and the Gobi.

Then, beginning about 140 million years ago, Earth underwent a further warming that marked the heyday of world climate. In these calm, equable conditions annual surface temperatures may have risen by as much as six degrees Celsius, spurring the growth of plant life dominated by the seed-bearing cycads and ferns that nourished the dinosaurs of the Jurassic and Cretaceous periods. Ice, sleet, and snow were unknown in this epoch of expanding seas and thunder-footed giants, whose fossilized relics we worship in cavernous ossuaries called museums.

Without warning, the curtain suddenly fell some 65 million years ago. Tyrannosaurus (which ruled the land), plesiosaurus (the sea), and archaeopteryx (the air) became part of a mass extinction triggered by a global disruption of the carbon cycle. The reason for this abrupt climatic change is the subject of debate among scientists. The leading hypothesis holds that a catastrophic event in the form of an asteroid impact was responsible, the evidence for which is provided by the great Chicxulub crater on the Yucatán Peninsula and in the adjacent Gulf of Mexico. The dust and smoke produced by the titanic collision greatly reduced the solar radiation required to heat Earth's surface, shrouding the planet in semidarkness. The flora withered and died for lack of photosynthesis; freezing temperatures enveloped the globe; the dinosaurs, whether warm-blooded or cold-, succumbed to starvation and hypothermia.

~

Nothing of this was known at the onset of the industrial revolution, nor was any aspect of the carbon cycle itself less understood than the part played by combustion, though Prometheus had long since stolen fire from the gods. In ancient Greece fire was considered one of the four basic elements, a substance from which all things are composed; in Rome the most carefully preserved cult was that of Vesta, goddess of the hearth, who together with her attending virgins guarded holy fire lest it go out and bring on a national calamity. The Zoroastrians of Iran placed fire at the center of their religion and regarded it as the most potent and sacred power ever given to man. The Aztecs of Mexico and the Incas of Peru worshiped gods of fire with sacred flames, which the Inca are said to have ignited by concentrating the Sun's rays with a concave metallic mirror. In India's Vedic scriptures Agni is the messenger between mortals and their gods, the personification of sacrificial burning.

The step from man's control of fire to its manufacture is huge and took hundreds of thousands of years. Hearths have been found in the Old World with the remains of the Neanderthals, who lived during the last ice age, between 100,000 and 30,000 years ago. But the archaeological record contains no clues as to how these fires were kindled, for the materials employed in artificial production are highly perishable. Not until the Neolithic period, when plants and animals were finally domesticated, is there physical evidence that humans actually knew how to produce fire at will.

Most did so via the friction method, using a pointed stick of hard wood and a piece of softer wood. The stick is rolled between the palms or spun by means of a thong and bow while it is pressed into a hollow on the edge of the soft material. The resulting heat produces glowing wood dust that ignites when gently blown upon and touched to small bits of kindling. With this simple technology came the ability to cook food; to clear land for agriculture; to provide warmth and illumination in hovels silhouetted against the long, forbidding night. In the magic that is fire, clay vessels became pottery; copper and tin were combined to make bronze; iron was forged into steel. In 1805, when the French chemist Gustav Chancel tipped cedar splints with a paste of chlorate of potash and sugar, creating the first "safety" matches, he dubbed his handiwork Prometheans.

Yet it was to counterbalance the gift of fire that Zeus sent the beautiful Pandora to Earth with her box of nasty surprises. With the rapid spread of controlled fire in the industrial age, the amount of CO_2 in the atmosphere began to increase. In 1850 the global CO_2 level of the atmosphere was roughly 280 parts per million (ppm), whereas by the mid-1990s it had increased to approximately 360 ppm. Moreover, this increase accounts for perhaps half the estimated amount of CO_2 entering the atmosphere. The remainder has largely been taken up and stored by the seas, which can absorb only a finite amount of the gas.

As Fourier correctly surmised, Earth's atmosphere is such that a portion of the Sun's rays are able to penetrate this delicate shield and reach the planet's surface, where they heat it. Some of this energy is reflected back into the atmosphere in the form of long-wave infrared radiation, much of which is absorbed by molecules of CO_2 and water vapor. This energy is then reflected back to Earth as heat. The process has been compared to the effect produced by the glass panes of a greenhouse, which transmit sunlight in the visible range but trap the heat. Were it not for the warming of Earth's surface and lower atmosphere, the planet's temperature would plummet by some fourteen to fifteen degrees Celsius (some 60°F), turning the oceans into ice and freezing the mammals that succeeded the dinosaurs in their tracks. However, as the level of CO_2 rises, the protective shield of gas thickens, trapping more heat, and global temperatures increase—or so many scientists have come to believe. Should temperatures rise by some two to three degrees Celsius (about 3.6° to 5.4°F) in the twenty-first century, the negative effect on human welfare, not to mention that of other species, could be devastating, as we shall see.

Such considerations could not have been farther from the minds of those who grew up during the Enlightenment and the early industrial age, when reason and material progress were believed to be synonymous. One of the best exemplars of this profound optimism was Philadelphia's Benjamin Franklin, political operative, inventor, and scientist par excellence.

During his first visit to England in 1726, the twenty-year-old Franklin became a fixture at Bateson's Coffee House, where he was introduced to the natural philosopher Henry Pemberton by a mutual friend. Pemberton had just seen the third edition of Isaac Newton's *Principia* through the press, and the editor promised Franklin a meeting with its eighty-three-year-old author, who had become an international icon. "This never happened," the crestfallen young American later wrote in his autobiography, and Newton was dead within a year.

If not Newton himself, then why not the great man's method? Following his return to the colonies, Franklin mastered the new experimental science with what one admirer termed "a head to conceive" and "a hand to carry into execution." After thinking long and hard on a problem, such as "electrical action," the gifted craftsman designed and constructed the instruments required to test his various hypotheses. In doing so he was the first American to establish an international reputation in pure science and the first natural philosopher to gain fame for work done wholly in electricity.

Less well known than his *Experiments and Observations on Electricity* is a long letter Franklin composed in August of 1785, when he was going on eighty and crossing the Atlantic for the last time on his way home from England. It was addressed to his "dear friend" Dr. Jan Ingenhousz, a Dutch national and court physician to the emperor at Vienna. More a minor treatise than a missive, the fifty-six-page document, enhanced by a copper plate of Franklin's design, was later published under the title *Observations on the Causes and Cure of Smoky Chimneys* (1787).

Having invented back in 1744 the Franklin, or Philadelphia, stove—an iron fireplace insert that furnished greater heat while conserving fuel—the author continues to be amazed at the general ignorance of how fireplaces and chimneys actually function. "Many are apt to think that smoke is in its nature and of itself specifically lighter than air, and rises for the same reason that cork rises in water." Franklin, whose mental acuity has not dimmed with age, then describes a number of experiments he conducted which prove that "smoke is really heavier than air; and that it is carried upwards only when attached to, or acted upon, by air that is heated, and thereby rarefied and rendered specifically lighter than the air in its neighborhood."

Case closed.

Franklin's physician friend might also be interested to know that the American has conquered his "aerophobia," or dread of cool air. "I considered it as an enemy, and closed with extreme care every crevice in the rooms I inhabited." Now he knows better and is persuaded that "no common air from without is so unwholesome as the air within a close room that has been often breathed and not changed." He works and sleeps with a window open and "can lie two hours in a bath twice a week . . . without catching a cold," his squat, cherubic frame bobbing in the breeze.

Franklin further observes that chimneys have been in general use in England only since Elizabethan times, before which each dwelling had only one location for a fire whose smoke exited through a hole in the roof. "Such [has been] the growth of luxury, that, now every possessor of a chamber, and almost every servant, will have a fire." He reports that the increased consumption of wood has resulted in the virtual disappearance of England's once magnificent forests and will soon render fuel extremely scarce in France unless its citizens do as the English have done and switch to coal.

At first, the transition to the fossil fuel was met with opposition among London's dyers, brewers, smiths, and other craftsmen, who attempted to have its burning outlawed by Parliament, arguing that it "filled the air with noxious vapors and smoke, very prejudicial to health, particularly of persons coming out of the country." Its use was resisted by householders as well, but economic forces took over and their resolve soon faded. Franklin the scientist forgets himself when he concludes: "Luckily the inhabitants of London have got over that objection; and now think it rather contributes to render their air salubrious, as they have had no general pestilential disorder since the use of coals, when, before it, such were frequent." If the French and the Germans fail to overcome this same "prejudice," it will be at their peril, for on coal rests the future of the civilized world—a prediction destined to be fulfilled far sooner than the sage of Philadelphia envisioned.

Part Two
THE WORLD EATERS

All my means are sane; my motive and my object mad.

—Herman Melville, *Moby Dick*

4

"QUEST FOR THE BLACK DIAMOND"

~

'Twould ring the bells of Heaven
The wildest peal for years,
If Parson lost his senses
And people came to theirs,
And he and they together
Knelt down with angry prayers
For tamed and shabby tigers
And dancing dogs and bears,
And wretched, blind pit ponies,
And little hunted hares.

—Ralph Hodgson, "The Bells of Heaven"

What Thomas Jordan, an English coal miner born in the reign of Queen Victoria, most remembered about his life below ground was the ponies. When they were old enough, the terrified beasts were trussed and lowered into the shaft, where they labored for years in the enveloping darkness, going blind for lack of exposure to natural light.

Thomas had not yet turned fourteen when he parted ways with school and, dressed up as a little miner in a blue flannel shirt, short breeches, and sturdy shoes, descended into the pit at nine o'clock on a cold January night. There he remained until eight the following morning, when he was relieved by another youth his age. During the intervening hours of nerve-racking pandemonium, punctuated by staccato shifts in the strata overhead, "the quest for the black diamond," Jordan later recounted in his unpublished autobiography, "overruled sanity, order and grace."

He began work as a "trapper-boy," whose primary responsibility consisted of opening and shutting a door to allow the "putters," or drivers, to pass through with their ponies and brimming coal tubs. Once emptied, the tubs were sent hurtling back down the shaft in groups of sixty, forcing every man in their path to dive into a series of strategically dug holes to avoid being crushed to death. Those who loaded the coal were paid by the tub, and Jordan never forgot the day when one of the manic laborers came running up to him stark naked, demanding more empty containers. He explained that he had just gotten married and needed to buy furniture for his bride.

Jordan was later promoted to driver and formed a bond with the ponies, which became his partners in danger. The roof of the pit was in deplorable condition, pocked by huge stones that might let loose at any moment. A pony frequently got hung up in the narrow, twisting shaft, its collar snagging on some object protruding from above. Freeing it was a tricky business and could be fatal if the beast suddenly panicked and began to flail. At other times the animal would turn too sharply, causing the tub it was pulling to strike one of the flimsy side props. When this happened, Jordan fled for fear of being buried beneath an avalanche of rock.

Once, during a prolonged labor strike, the ponies were hoisted to the surface and allowed to roam free across the green fields of Durham, where they nibbled fresh shoots and rolled on the warm pasture for the first time in years. Jordan was so moved he wept for joy. Then the strike was settled, and the pulley wheels began to turn again. The miner's heart broke as he watched the ponies being rounded up and lowered back into the abyss. To this sensitive man of literal faith and simple metaphors, the mine was a demonic prison, "a hell condemned to," just as the village green where he strolled after work was a heaven, the nearby River Wear a halo whose waters cleansed his aching body and restored his battered spirit.

Bad as it was, Thomas Jordan's forebears had endured far worse. In the eighteenth century, many of those working in England's mines were bound to their owners by the year, a condition approximating serfdom. The anonymous author of a scathing pamphlet published in 1756, who may have been Edmund Burke, the trenchant political philosopher and government critic, declared: "I suppose that there are in Great Britain upwards of a hundred thousand people employed in lead, tin, iron, copper, and coal mines; these unhappy wretches scarcely ever see the light of the sun; they work at a severe and dismal task, without the least prospect of being delivered from it; they subsist upon the coarsest and worst sort of fare; they have their health miserably impaired, and their lives cut short by being perpetually confined in the close vapor of these malignant minerals."

The state of technology was such that the coal shafts were as yet too narrow to accommodate four-legged beasts of burden. Much of the hauling was done by children as young as six, called "little trappers," and women, some stripped to the waist, others completely naked, who trailed baskets or pushed rude sledges through the claustrophobic darkness on scarred hands and knees. In addition to soaking in their own perspiration, many were perpetually bathed in the muddy pools formed by leaking ceilings and dripping walls, for England's high water table made flooding a perpetual danger. In an attempt to retard the flow, the shafts were lined with sheepskins, while bucket brigades and crude hand pumps were regularly called into action. Upon returning home in the evening, their faces unrecognizable beneath a long day's accumulation of grime, the young women again undressed to the waist and washed themselves in a tub of lukewarm water, not bothering to close the cottage door but chatting amiably with any young man who happened by. Here matters might have stood had not an unforeseen conjunction of circumstances, not the least of which was the climate, catapulted England into the modern industrial age.

~

Some 12,000 years after the glaciers retreated, Europe experienced a climatic change reminiscent enough of the Pleistocene epoch to be called the Little Ice Age. But before its advent in the fourteenth century, Europe's climate was marked by warm temperatures and moderate rainfall, which spurred the cutting of forests and the extension of agriculture, especially the development of English vineyards. After some two centuries of comparatively halcyon climes, the warmer years gave way to cooler ones; gentle spring rains became winter deluges. Between the Renaissance and the nineteenth century, mean temperatures were probably only a degree or two Celsius below those of the present day. Yet on January 30, 1649, when Charles I met his God on the scaffold erected before White-hall, huge chunks of ice clogged the arches of London Bridge, causing the monarch to don two sets of underclothes lest he shiver, thus giving his enemies the ultimate triumph of think-ing him afraid. The same ice-choked Thames froze solid in Dickens's time, inspiring the novelist's colorful portrayal of holiday skaters, who have long since vacated the scene. Clima-tologists with a historical eye have scrutinized many other cre-ative works, most particularly paintings, in which drifted snow, daggerlike icicles, and heavy clothing lend substance to T. S. Eliot's observation that "April is the cruelest month," or so it must have seemed little more than a century ago.

The combination of a cooler climate, a rising population, and the desire of landowners to profit from greater agricul-tural production was first felt in London, where the price of wood transported over ever greater distances was on the in-crease. Still, relatively little coal was burned in the city during the late Middle Ages and Renaissance, owing to poorly de-signed fireplaces and chimneys. Londoners preferred smoke-less charcoal to coal. The charcoal was prepared by piling wood into stacks, covering it with earth or turf, then setting it afire. During combustion, the volatile compounds in the wood pass off as vapors while the remaining carbon is con-verted into charcoal. By weight, charcoal gives more heat than wood and, lacking branches, is easier to pack into carts. The advantages of a smokeless hearth were largely offset by the pollution resulting from the production of the increasingly popular fuel. Large smoky kilns were built ever nearer to the city, darkening the overhead sky and blackening the exteriors of the architectural monuments below, not to mention the homes of the prosperous and humble alike.

The drenching winter rains of the Little Ice Age rendered England's roads all but impassable to heavy carts and wagons at the very time of year when the demand for fuel was at its peak. Drivers in command of the strongest teams of draft horses and bullocks were often stopped in their tracks, while the lone horseman watched helplessly as his struggling mount sank to its belly in the muck. This, in addition to a continuing rise in the cost of wood, forced authorities to seek an alternative energy form as well a cheaper and more efficient means of transporting it to the capital.

They eventually found it in "sea-coal," so named because it was carried by ship from the northern ports nearest the mines to docks on the lower Thames. A small trade in the fuel had actually been going on since Tudor times, but as previously noted, its burning was thought detrimental to the general health. A well-bred Elizabethan lady would not so much as set foot in a room where sea-coal was being burned, let alone consume food prepared over such a fire. Nor would her no less fastidious spouse take beer tainted by the smell of coal smoke.

It so happened that Scottish nobles knew something about coal of which their English counterparts were ignorant. When, in 1603, an uncompromising James VI of Scotland succeeded Queen Elizabeth, becoming James I of England, he ordered that coal be burned in the fireplaces of his household in London. Instead of the sulfurous and smoky type mined in northern England, known today as bituminous, he ordered the importation of hard, clean-burning anthracite from his native Scotland. James's eager-to-please courtiers followed their king's example, helping to set the stage for the eventual triumph of the black diamond.

By the beginning of the eighteenth century, London had made the transition from a wood-burning city to one that relied mainly on imported coal. Owing to the increased height and efficiency of chimneys, most households consumed the cheaper bituminous, which burned with a smoky yellow flame. Production in the mines of Lancashire, Durham, Cornwall, and elsewhere rose exponentially, spurring the opening of new collieries as rapidly as human muscle would allow.

As always, the problem of underground seepage continued to plague owners and miners alike, cutting into profits, threatening jobs, and, in some instances, claiming human lives without warning, usually where shafts were sunk along the coastline and extended under the sea. In 1695 Celia Fiennes, a dauntless traveler and diarist, noted that in the West Country, "they even work on the Lord's day to keep the mines drained—one thousand men and boys working on drainage of twenty mines." Even so, the sea frequently won as dozens of mines were lost to the overwhelming waters.

Among the observers of this seemingly endless waste was Thomas Newcomen, an ironmonger from Dartmouth and descendant of an aristocratic family that had lost its property in the wake of Henry VIII's connubial machinations. In the 1680s Newcomen formed a partnership with John Calley, a plumber and fellow Baptist. Together they made the rounds of the local mines, supplying their owners with metal and undertaking work on site. After retracing the mostly unsuccessful footsteps of others during years of trials and tinkering, Newcomen, with Calley's assistance, built the first successful steam engine in 1712. It was designed for one purpose only—the removal of water from inundated mines.

The device was an ingenious combination of what would soon become familiar elements of the nascent industrial age: a piston and cylinder, pumps, valves, levers, and, most creative of all, a process of producing low pressure by the condensation of steam in a vessel. Newcomen had accidentally discovered that a small jet of cold water introduced into the single piston as it reaches a full head of steam causes the resulting air pressure to push the piston down, which in turn raises the suction pump rod, lifting the ground water to the surface. Within weeks, he had redesigned his engine so that this "accident" occurred precisely once every stroke.

The Newcomen engine was constructed directly above the mine shaft, its giant metal cylinder encased in a circular brick boiler house that towered over the men who kept it stoked. It was extremely inefficient; some of the largest engines may have taken a load of fuel in excess of thirty tons. That mattered little, however, for coal was the one resource the rejuvenated mines contained in plenty. A decade after its unveiling, the steam engine was in operation all over Europe, making its contribution to a burgeoning economy while adding to the pollution of what some had already begun to refer to as the "coal age."

~

In developments that at the time seemed to have nothing to do with Newcomen's engine and the increased mining of coal, Britain's textile production was being revolutionized by a series of inventions about to link its fate, and that of modern society, to carbon and steam—and therefore to the greenhouse effect. This part of the story begins with John Kay, the namesake of a Colchester woolen manufacturer. In 1733 the younger Kay, who had recently marked his twenty-ninth birthday, took out patent No. 542, which set forth what was perhaps the most significant improvement ever made in the loom.

Up to that time the shuttle, a wooden holder that carries the weft, or crosswise threads, through the alternating vertical fibers of the warp, was thrown by the weaver with one hand and caught by the other. Since most looms of the period were designed to weave broad fabrics and exceeded the span of a man's arms, two weavers were required to throw the shuttle back and forth. Kay transformed this inefficient process by designing what he at first called his "wheel-shuttle" loom. Instead of tossing the shuttle by hand, the weaver had only to trip a cord attached to a spring, which in turn sent the wheeled shuttle speeding between the warp threads automatically. Not only did the new loom work much faster than the older ones, but broadcloth could now be woven by a single weaver rather than two, more than doubling production. Little wonder that the fascinating device was soon rechristened the "flying-shuttle."

Unfortunately for Kay, the invention that should have made his fortune nearly turned him into a bankrupt. Although the new loom was quickly adopted by the woolen manufacturers of Yorkshire, they defied the law by refusing to pay its inventor the royalties due him. Kay's supporters responded by forming an association called the Shuttle Club to help defray the expenses incurred during the years of litigation that followed.

Humiliating and frustrating though this was, Kay was about to suffer an even greater indignity. Prior to his invention, it is estimated that it already took ten spinners to keep a single weaver supplied with yarn, an imbalance that had silenced many a loom. With the advent of the even more efficient shuttle, the situation became intolerable. In 1745 an angry mob broke into Kay's house in Bury and smashed everything they could lay their hands on. Kay himself might have been seriously injured, perhaps even killed, had not his friends wrapped him in a wool sheet and slipped him through

the ranks of the trespassers, who mistook them for fellow loot-
ers. Taking no chances, Kay withdrew to France, where he later
died, obscure and all but penniless.

The acute shortage of yarn continued to plague the textile
trade for another generation. Then, in 1764, a stout, muscular
carpenter and hand-loom weaver from Blackburn named
James Hargreaves had a sudden insight, perhaps after observing an ordinary spinning wheel overturned on the ground, its
normally horizontal spindle in an upright position. Hargreaves set about crafting a machine with eight vertical spindles
located side by side, each capable of holding its own thread.
He named his creation the spinning jenny, after his wife, and
warily kept his invention secret, using it only to make weft for
his own loom. Having many mouths to feed, Hargreaves eventually constructed additional machines, adding extra spindles
as he refined his original design. These were sold to other
spinners, who quickly discovered that the small, agile fingers
of children were most adept at working the multiple threads.

In the spring of 1768, twenty-three years after John Kay
had escaped a violent mob, Hargreaves's own house was gut-
ted, his jenny and loom smashed by old-fashioned spinners
who feared for their livelihood. The inventor migrated to Nottingham, where he formed a partnership with a woodworker,
and together they built a small cotton factory in which the
jenny was utilized. In July of 1770, Hargreaves took out patent
No. 962, titled "Abridgments of Specifications for Spinning,"
then promptly entered the legal lists to protect the rights of his
brainchild from the same unscrupulous manufacturers who
had copied the flying-shuttle without paying Kay his due.

At the time of Hargreaves's death, in 1778, a single spinner
was able to manage as many as a hundred or more spindles at
a time. The threads were not strong enough, however, to be
used for the warp in cotton cloth, necessitating further refine-
ments in the spinning process.

As the youngest of some thirteen children from an impoverished Lancashire family, Richard Arkwright hardly seemed a
promising candidate for invention. About all that is known of
his early life is that he was apprenticed to a barber and attended school during the winter months. He somehow became acquainted with the mechanics of textile production

Richard Arkwright
(Used by permission of The Granger Collection, New York)

and was drawn to the problem of how to mass-produce warp thread of a strength greater than the softer weft fibers spun on Hargreaves's jenny. It occurred to him that spinning rollers, which had already been tried by others with limited success, might hold the key.

Together with two friends who supplied the funds and some much-needed expertise, Arkwright set up operations in 1768 in the parlor of a house rented from the Free Grammar School of Preston, chosen, it was said, because it was well protected by a garden of prickly gooseberries. Being poorly dressed and a stranger, the secretive Arkwright became an object of suspicion in the eyes of the locals. Rumors of sorcery and witchcraft began to circulate, supported by the accounts of two elderly women who lived nearby. Many strange noises, especially of a humming nature, could be heard at all hours, as if the Devil were tuning his bagpipes while his disciples danced a reel. Perhaps the men of Preston should do their Christian duty and break down the door.

Having satisfied himself that his invention was indeed practical, Arkwright discreetly headed for the friendlier confines of Nottingham, in 1769, where a victimized Hargreaves had settled the year before. With money ventured by others, he erected a cotton mill and took out a patent for the machine he called the spinning frame. As the cotton passed between a series of four rollers, it was drawn to the requisite fineness for twisting, which was accomplished by spindles, or flyers, placed in front of each set of rollers. The resulting thread was hard and suitable for warps, whereas that spun on the jenny was still superior for wefts.

Arkwright's spinning frames were at first propelled by horses, but the animals were expensive to purchase and costly to maintain. They were also incapable of generating the power required for ever larger machines. After securing additional partners, the resourceful entrepreneur moved his operations from Nottingham to Cromford on the Derwent. In this deep, picturesque valley, where he could command the river's flow, Akwright renamed his invention the water frame and undertook the manufacture of bright and much-sought-after calicoes. Additional patents followed, as did the construction of even bigger mills until, like the Rockefellers, Carnegies, and Fords of the future, Arkwright was able to fix the price of his product, forcing others to follow suit or fold their tents.

Arkwright now looked like a Dutch burgher in a Rembrandt oil; his face, once gaunt, had grown pudgy, giving him an eternally youthful aspect, while his spare frame thickened into obesity, the gold buttons of his waistcoat straining to contain his creeping girth. It is from the building of the spinning mills of Arkwright that we mark the beginning of the modern factory system and the emergence of a new breed of wealthy capitalists.

~

Arkwright's onetime neighbors were not the only ones fearful of ghosts in the machine. A young and reclusive Samuel Crompton earned a scant living playing his homemade fiddle by day so that he could work on his latest invention by night. It was the sounds of his tinkering, combined with the flickering of midnight candles, that caused wary observers to believe that the Crompton place was surely haunted.

Samuel, his widowed mother, and his brothers and sisters occupied the caretaker's rooms of an ancient mansion at Firwood, near Bolton, called Hall-in-the-Wood. Having grown up spinning yarn which he wove into quilting, the youth was well acquainted with the imperfections of the spinning jenny, such as its irritable tendency to cause the yarn to break from too much tension. For five years, during his twenties, Crompton labored in secret over his new machine, which he finally completed in 1779 and christened Hall-in-the-Wood, after his birthplace.

In truth, Compton had succeeded in building an ingenious hybrid that combined the principle of Arkwright's rollers with the spindles of Hargreaves's jenny, but with an invaluable addition. He added what he termed the "moveable spindle carriage," which alleviated the strain on the thread. The process was now under the control of the spinner, who regulated the tension by deft movements of the hand and knee. The result was a much finer yet stronger thread than that produced by either the water frame or the jenny. Moreover, Crompton's machine could spin yarn for both warps and wefts.

Crompton's neighbors, their fear overcome by curiosity, risked injuring themselves by climbing up to his second-floor window to get a glimpse of the mysterious invention. Its builder countered by erecting a screen to hide himself, but the annoyance got so bad that it interfered with his work. Never comfortable in a social setting, Crompton longed for a return to anonymity. In exchange for the vague promise of a small royalty from manufacturers, he committed Hall-in-the-Wood to the public domain, where it was soon renamed the mule for its hybrid origins. Only in 1812, some thirty-three years later, did Crompton receive a parliamentary grant of £5,000. By then the mule's 5 million spindles were providing employment to 70,000 spinners directly, and indirectly to 150,000 weavers, while Hargreaves's jenny and Arkwright's water frame were rapidly becoming the first industrial relics.

It fell to a poet and cleric with two degrees from Oxford to complete what the others had started. After taking holy orders on leaving the university, Edmund Cartwright became a curate at Brampton, where he settled into the life of a country parson, his financial position secured by marriage to an heiress. In 1722 he published anonymously *Armine and Elvira*, an epic poem that went through numerous editions and won the praise of Sir Walter Scott.

In the summer of 1784, during a visit to Matlock near the site of Arkwright's mills at Cromford, Cartwright entered into a conversation with some gentlemen from Manchester. One of them observed that the patent on Arkwright's spinning machine would soon expire; when it did, so many mills would be erected and so much cotton spun that there would never be enough hands to weave it into cloth. Cartwright replied that the great inventor must now set his wits on building a weaving machine to equal his water frame. This comment was met with derision. Such a thing was impracticable; besides, Cartwright, by his own admission, had never so much as seen a weaver at work. The wounded parson tried to defend himself by arguing that an automatic chess player had recently been exhibited in London. "Now will you assert, gentlemen, that it is more difficult to construct a machine that shall weave than one which shall make all the variety of moves which are required in that complicated game?"

The memory of this unpleasant encounter lingered until Cartwright finally hired a carpenter and smith to build a device that incorporated the three successive movements he envisioned in the weaving process. Never having turned his thoughts to anything mechanical before, he was forced to admit that his creation was "a most rude piece of machinery." And yet it worked, albeit laboriously. "The warp was placed perpendicularly," he recalled, "the reed fell with the weight of at least half a hundredweight, and the springs which threw the shuttle were strong enough to have thrown a Congreve rocket. In short it required the strength of two powerful men to work the machine at a slow rate, and only for a short time."

Nevertheless, Cartwright patented his ugly duckling in 1785, "considering in my simplicity that I had accomplished all that was required."

Not quite. Some months later the first-time inventor "condescended" to observe a hand weaver at his loom. Cartwright was "astonish[ed] when I compared their easy modes of operations with mine." He immediately set about refining his invention until, two years later, he obtained a second patent on a power loom that did not differ substantially from the thousands of others that would soon follow.

As with Kay and Hargreaves, there was a stiff price to be

paid. Cartwright first set up a spinning and weaving factory at Doncaster, where he continued to experiment. His large loom, driven by a bull, doubled the size of the cloth even as it greatly speeded its production. In 1791 a cotton firm in Manchester contracted with him to build a factory large enough to hold 500 power looms, some to be driven by a revolutionary form of the steam engine. Once in operation, the looms cost only half the wages normally paid to hand weavers, and it was not long before angry threats were being made. Shortly thereafter the new mill was set afire and burned to the ground, ruining Cartwright, who had sunk most of his wife's fortune into his invention.

Not until 1809 did Parliament vote the power loom's inventor £10,000 for building one of the machines that made the industrial revolution possible. The retired cleric bought a small farm at Hollander, where he passed the rest of his life writing middling poetry and designing agricultural equipment. In his eighty-third year he sent a paper containing a new theory of the movement of the planets around the Sun to the Royal Society, which declined to publish it on the grounds that Sir Isaac Newton had already explained the phenomenon quite well enough.

~

As the eighteenth century drew to an end, the textile industry stood poised on the threshold of the industrial era. Robert Frost, whose poetic gifts far exceeded those of Cartwright, pictured the scene created by the men who had labored to replace hand tools with modern machines:

> The air was full of dust of wool.
> A thousand yarns were under pull,
> But pull so slow, with such a twist ,
> All day from spool to lesser spool,
> It seldom overtaxed their strength;
> They safely grew in slender length.
> And if one broke by any chance,
> The spinner saw it at a glance,
> The spinner still was there to spin.
> That's where the human still came in.

The cadenced hum of the looms and spinning machines were still dependent on the flow of the river, so it followed that the new factories and mills were mostly built in country districts. That was about to change, however, due primarily to the efforts of an obscure instrument maker and repairman who encountered one of Thomas Newcomen's steam engines forty-one years after its invention.

Born in Greenock, Scotland, in 1736, James Watt was descended on his mother's side from a family that had once been prominent in Scottish life. His father was a shipwright and a supplier of nautical instruments, many of which he made himself. Owing to fragile health, Watt's attendance at school was somewhat irregular, and it was in his father's workshop that he acquired the skills of a master craftsman. Later on, with the aid of friends on the faculty of the University of Glasgow, he gained an appointment as mathematical instrument maker to the university. It was during the course of his duties that he was asked to repair a model of the Newcomen engine.

While engaged in this task, he realized that the machine was extremely inefficient. Though the jet of water condensed the steam in the cylinder very quickly, it had the undesirable effect of cooling the cylinder down, resulting in premature condensation on the next stroke. In effect, the cylinder had to perform two contradictory functions at once: it had to be boiling hot in order to prevent the steam from condensing too early and also be cold in order to condense the steam at just the right time. Watt redesigned the engine by adding a separate condenser, allowing him to keep the cylinder hot by jacketing it in water supplied by the boiler. The result was an immensely more powerful machine than the Newcomen engine, which was essentially little more than a giant pump. The inventor applied for and received a patent in 1769.

Watt's initial success was followed by a remarkable sequence of improvements of his own making. The most important of these was the sun-and-planet gearing system, which translated the engine's reciprocating motion into rotary motion. In simple terms, the new machine could be used to drive other machines, and Watt calculated its mechanical muscle in units he called horsepower.

Unlike Arkwright, Watt lacked financial savvy. He once admitted that he would "rather face a loaded cannon than settle a disputed account or make a bargain." He closed his shop at the university and, after opening a marginal land surveying and civil engineering office in Glasgow, moved to Birmingham, where he fell into debt attempting to market his engine. His financial straits were made even more perilous by the endless patent litigation that seemed to plague every British inventor worth his salt. Watt was about to declare bankruptcy when he was rescued by Matthew Boulton, a wealthy Birmingham hardware manufacturer. With Boulton providing the capital, the two men formed a partnership, and by 1800 the firm had sold 289 engines for use in factories and mines.

Watt alone had turned a steam pump into a machine that both spurred and drove the industrial revolution. As early as 1787, its adaptability to a variety of uses was described as "almost incredible," while *European Magazine* stated that it would change "the appearance of the civilized world." From now on the new factories, both textile and iron, would be constructed in the cities, forever altering the worker's way of life.

It was at this dual moment of triumph and anticipation that Watt received a somber letter from one Dr. Percival, a prominent Manchester physician. Percival wanted to know if Watt was privy to any methods of abating the smoke nuisance caused by the escalating consumption of coal. Watt had no solution to offer, nor could he foresee the day when his own steam engine would greatly magnify the problem.

Most women and very young children were soon replaced in the mines by the pit ponies that would also disappear once the steam-driven coal cutter was introduced underground. And like the effluents of industry, the cryptic moth was taking to the skies. In 1860 a specimen of *carbonaria* was captured in Cheshire; in 1861 another was netted in Yorkshire; in 1870 Westmorland recorded its first black mutant; and in 1878 Staffordshire was added to the list. Nor was the continent far behind. In 1867 a mating pair was captured on an elm tree in the Netherlands. Seventeen years later, in 1884, their descendants had made it to Hanover, and in 1888 to Thuringia and up the Rhine Valley, conquering Europe wherever the sky was blackened with soot. At rustic Down House, where the air remained clear, not a dark wing fluttered in Darwin's peaceful garden.

5

CLEOPATRA'S NEEDLES

Above the smoke and stir of this dim spot
Which men call earth.

—John Milton, "Il Penseroso"

In the 1870s, well before he left Europe for the United States and, later, the South Seas, a gaunt but ruddy-faced Robert Louis Stevenson wandered the length of France, occasionally in the company of a recalcitrant donkey he christened Modestine, which he compared in size to a large Newfoundland dog and in color to "an ideal mouse." In his first travel book, *An Inland Voyage*, the author wrote of entering the town of Noyon on the Oise, where he was drawn to the straight-backed Cathedral of Notre-Dame with its two stiff towers that rose solemnly above the tile roofs of the tightly clustered houses. "I find I never weary of great churches," Stevenson mused, "it is my favorite kind of mountain scenery. Mankind was never so happily inspired as when it made a cathedral; a thing as single and specious as a statue to the first glance, and yet, on examination, as lively and interesting as a forest in detail. The height of spires cannot be taken by trigonometry; they measure absurdly short, but how tall they are to the admiring eye!"

Happily inspired, perhaps, yet it was by "trigonometry" that medieval Europeans ranked the soaring edifices in their midst: The size of a cathedral and the height of its vaults and spires meant everything to rivals for the lucrative pilgrimage trade and, beyond that, the blessings of an ever watchful God.

Even as Stevenson was recording these thoughts, brave masons of a different sort were topping off much taller structures across Britain. Known as "the tall chimneys," these were the soaring industrial colossi of the Victorian age, shaped from brick, stone, iron, and concrete, rising so high that their builders were sometimes enveloped by low-hanging clouds, as if in apotheosis.

In 1885 Robert M. Bancroft, onetime president of the Civil and Mechanical Engineers' Society headquartered in London, and his son, Francis, an assistant municipal surveyor, published a fascinating book with the pedestrian title *Tall Chimney Construction*. Within its pages not only did the Bancrofts present the details of eighty industrial chimneys built in England, the United States, and on the Continent between 1835 and 1883, but they covered the subject from the laying of foundations to the installation of lightning rods, omitting little in between.

The opening paragraph presents the dual rationale for tall chimney construction: "Firstly, to create the necessary draught for the combustion of fuel; secondly, to convey the noxious gases to such a height that they shall be so intermingled with the atmosphere as not to be injurious to health." Regarding the second objective, the Bancrofts have nothing further to say, since they subscribed to the conventional wisdom that industrial smoke released from a sufficient altitude (a minimum of 90 feet according to the Town Improvement Acts of Manchester, Bradford, and Leeds) would soon disappear, leaving few traces other than blackened capstones. Of the first purpose, they write a great deal more.

Harkening back to Benjamin Franklin, the authors explain how the column of heated air in a chimney is lighter than the corresponding column of "atmospheric air," which presses into the entrance of the furnace, causing the heated air to rise. The normal air is in turn heated and displaced, creating a natural draft. In engineering terms, the column of cooler air and the rarefied air in the chimney are comparable to a set of scales, or the two ends of a lever of which the boiler is the fulcrum. Thanks to Watt's perfection of the steam engine, the tendency of heated air to rise could be increased manyfold, first by placing powerful mechanical blowers between the boilers and chimney flue, and second by erecting taller chimneys, so that the difference between the volumes of heated and unheated air is made greater in proportion to the increase in height. The inrushing air also causes the coal to burn at a higher temperature, conserving fuel while making it possible to employ cheaper, high-sulfur bituminous, commonly known as "slack." In sum, an engineer operating in this new competitive climate could derive greater power for his engine while generating more steam. Yet the advantages of increased efficiency and economy carried a price. The level of poisonous fumes and gases rose dramatically even as they were cast to the deceptively forgiving winds.

Few today are aware that industrial chimneys of the 1900s were at their tallest near midcentury, a time when Britain held a dominant technological lead over the rest of the world. Although figures are far from complete, an average of 100 chimneys per year were erected in London proper between 1846 and 1853. In the latter year alone 123 new stacks appeared on the already surreal horizon, adding weight to Shelley's trenchant observation that "Hell is a city much like London—a populous and smoky

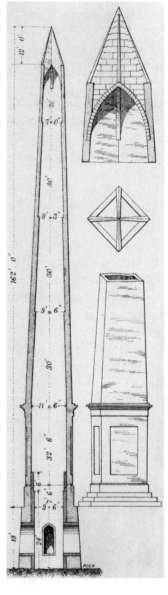

**Diagrams of
Cleopatra's Needle**
(Used by permission of the Hagley
Museum and Library)

city." Small wonder that as early as 1843 the House of Commons Select Committee on the Smoke Nuisance recommended that all manufacturers be removed to a distance of five or six miles from the city center.

Among the most critical elements of tall chimney construction was the invisible foundation, which was built on solid rock whenever possible. In many instances, however, chimneys had to be erected near rivers on sites where soft, alluvial clay made it necessary to excavate to a depth of thirty feet or more. In such conditions the foundation was usually made of concrete poured over pilings that were anchored in hard sand or bedrock. It was recommended that the finished foundation be allowed to cure for at least a month before commencing work on the superstructure.

Architects and factory engineers employed both brick and stone in chimneys reaching well over 300 feet, but by midcentury brick came to dominate for a number of reasons. It was uniform, strong, and easier to lay than stone. Moreover, brick lent itself to a number of architectural forms, whether round, square, or octagonal. Finally, brick was both easy to obtain and cheaper than most other materials.

Chimney construction began in late spring and ended in late summer to avoid the risk of frost, which could seriously damage the slow-setting lime mortar, threatening collapse. The bricks themselves had to withstand temperatures of 600°F and were laid at the rate of a 1,000 or more a day, an addition of 3 to 5 feet in height depending on the circumference of the shaft. It was critical that the shaft be plumbed and leveled every 3 feet to keep the stack erect, and each bricklayer was expected to keep an eye on the work of his fellow masons.

A chimney in excess of 300 feet required anywhere from 500,000 to over a million bricks, and it took a trio of masons up to three summers to lay them. In Lancashire large shafts were raised to about half their ultimate height, then left for six months to settle before completion. Even so, the slow-drying mortar kept the structure at risk during the first few winters. The effects of the prevailing winds meant that the mortar did not set at the same rate on all sides, subjecting the shaft to potentially disastrous oscillations. This danger was compounded by the expansion caused when heated gases began to pass upward through the flue. The problem was addressed by adding a lining of firebrick erected independently of the load-bearing walls. This material had to withstand gases as hot as 1,200°F and followed the shaft upward at least two-thirds of the way to the top.

In the first decades of the nineteenth century, exterior scaffolding surrounded the rising shafts, much as it had during the heyday of cathedral construction. But as chimneys grew taller, the traditional method proved too costly and such scaffolding was abandoned. It was replaced by a wooden mason's platform that spanned the opening and was held fast by timbers fitted into holes in the lining. Each time the platform was raised, the holes were bricked in as the workmen inched their way into the blue. Bricks, mortar, water, timber, and other materials were hauled up from the inside, sometimes with the aid of a steam winding engine, more commonly via a heavy rope and drum worked by common laborers below.

The first chimney to exceed the 400-foot mark was erected in Glasgow at the St. Rollox Chemical Works, in 1841–42. Rising 435.5 feet above the ground, the tapered shaft rested atop a poured foundation on the site of an old quarry, enabling the architects to dispense with the usual pilings. One and a quarter million bricks were laid in the span of a single year, a prodigious feat considering that work had to be suspended during autumn and winter. Indeed, a roof built over the unfinished spire to protect it from the elements was blown down in a December gale, leading to fears that the curing mortar might be permanently weakened. On June 29, 1842, the cope was laid, one year to the day after the foundation had been poured. Invisible to the assembled dignitaries and curious public were eighty acres honeycombed by underground flues extending to all parts of the sprawling factory.

How long the new record would stand was anyone's guess, yet few doubted that it was destined to be broken. In 1857 Joseph Townsend, owner of the Port Dundas Chemical Works, also located in Glasgow, signed contracts for the erection of a chimney that would eclipse that of his rival, St. Rollox. The height from ground level to the top of the coping ultimately measured 454 feet, a world record for industrial chimneys of the nineteenth century. By comparison, the Great Pyramid of Khufu, or Cheops, one of the Seven Wonders of the Ancient World, was slightly taller at 482 feet, but to reach this mesmerizing altitude countless laborers had employed a solid mass of limestone blocks covering thirteen acres. Townsend's 8,000-ton circular chimney, like most others of its kind, was freestanding and rested on compacted clay, with an outside diameter at its base of only 32 feet, tapering to just over 13 feet at the crown. Once the chimney was operative, its acres of branching flues carried off a variety of gases generated by the burning of bones and old horses for fertilizer, as well as the poisonous smoke associated with normal chemical operations.

Tall chimney construction was still a new and largely unproven technology when the industrial revolution came into its own. An inherently dangerous enterprise, it was capable of bringing both architect and engineer to grief.

At Newland's Mills in Bradford, an octagonal stone shaft rising 240 feet above the city was completed in 1863, but not before the structure developed a bulge that workmen repaired with iron wedges, thin stones, and mastic injected with syringes. Tenants of the mill owners became uneasy when, in 1882, a second bulge became visible from the ground, and the attempt to fix it failed. On December 27 a heavy gale blew the scaffolding down, along with pieces of the outer casing. The next night more of the casing fell. Suddenly the chimney, which had been built over an old coal shaft, began to settle. Seconds later the upper portion collapsed onto the crowded workers' quarters below, killing fifty-four people and destroying property worth an estimated £20,000. The legal action brought by the injured, orphaned, and widowed resulted in damages against the mill owners totaling £2,500.

Such a disaster was commonly known as "a fall," many of which occurred either during or shortly after construction. In June of 1873, a 220-foot stack was begun at the cement works of J. C. Gostling and Company, near Gravesend. The weather was good, and the chimney, built of the highest-quality brick and mortar, was completed in only sixteen weeks. On October 2 Gostling himself visited the site to witness the laying of the last brick. One man had already reached the top when the upper part of the shaft began to bulge. Immediately afterward the top 60 feet gave way, raining down bricks on the workmen below, six of whom were killed and another eight injured. The site was soon cleared, and the chimney was rebuilt the following year according to the original design.

The very wind counted on to disperse the effluents proved to be the construction engineer's biggest nightmare—the more so on an island buffeted by great storms sweeping in off the Atlantic. Although formulas were devised for the purpose of compiling tables on wind resistance, no one knew for certain how much lateral pressure a given structure could withstand. Obviously, a round shaft was superior to a square one when it came to meeting a gale head-on, but shape in and of itself was no guarantee of survival. Besides falling, chimneys were prone to tilt, threatening whatever or whoever happened to be below.

If the entire shaft was off-center, engineers first attempted to resolve the problem by attacking the foundation on the side opposite the list. Like prehistoric builders wrestling with a giant monolith, they removed the earth in stages, hoping to right the shaft gradually until it reached vertical once again. If this proved unsuccessful, they sometimes dug a well on the higher side of the shaft, then filled it with water in hopes that the sodden ground would cause the chimney to shift. Once it began to move, however, there was no way of controlling it; the structure was likely to reach vertical, only to begin tilting in the opposite direction.

The most reliable solution involved the use of screw jacks and iron plates. A gap was cut in the brickwork opposite the lean; then several bricks were removed and a jack was inserted together with metal plates at top and bottom. The process was repeated until several jacks were in place and the weight of the shaft had settled back onto them. With gravity now their ally, the workmen adjusted the regulating screws, causing the shaft to return to its original position. The sawn gaps were filled in with mortar; the screws were removed, one after another; and mortar was also put in their place.

A more common problem involved the straightening of a chimney whose base remained upright but whose middle and upper regions had begun to tilt, because of either wind damage or poorly cured mortar, sometimes both. Such was the fate of the world's tallest chimney, which doubtless would have fallen had not drastic action been taken by its Glasgow designer and builder, Robert Corbett. Struck by a northeast gale on September 9, 1858, the Port Dundas chimney began to sway. The next morning, its top was found to have shifted nearly 8 feet from vertical, the deflection beginning 100 to 150 feet above the ground. To save it, Corbett, with Townsend's acquiescence and workmen, gave the order to begin what engineers termed "sawing back."

Utilizing the interior scaffolding, which by good fortune was still in place, they punched holes through the multiple layers of brick with hammers and chisels, attacking the side opposite the tilt. Once this was done, they inserted heavy masonry saws and began cutting along the same grout line in opposite directions, four sawing and two wetting the searing blades as blister formed upon blister. Eventually, they were able to remove a partial row of bricks and replace it with a thinner course, filling the remaining space with iron wedges. They began at the 41-foot level, working their way toward the top in 30- to 40-foot increments. The exhausting task took days, and tensions, which remained high, peaked as the twelfth and final cut was completed 326 feet above the city. While Townsend fretted and paced below, his spent men inserted the last trowel of mortar and gradually made their way back down, removing the wedges as they descended. Within hours the stack had settled upright.

An elated Townsend opened the site to a captivated public, which he invited to visit the top. So great was the turnout that people by the hundreds waited as much as half a day to ascend the monster chimney, memories of which were handed down from one generation to the next like family heirlooms.

Although his chimney was not nearly as tall, the owner of the Stanton Iron Works Company outside Nottingham did Townsend one better. When the 190-foot shaft was all but finished, he provided a hot dinner for forty-seven of the workmen at the top. Three young women ascended to lay the tablecloth and wait upon the guests of honor, an event that caught the attention of the London press. Company board members soon began joining the laborers at the top of the new stacks, occasionally replacing them altogether. When this happened, the masons celebrated on their own by polishing off a barrel of dark ale.

With its furnace serving as a giant maw, the tall chimney spewed tons of pollutants into the atmosphere round the clock, its boilers shutting down only for occasional inspections and repairs. At the New York Steam Heating Company on Greenwich Street, the 1,000 tons of coal consumed each day were delivered by a primitive elevator, enabling four stories of boilers to generate a combined 16,000 horsepower. Next to the logistics of coal delivery, the biggest problem was what to do with the residual ash. The debris had to be cleared from the furnace with an iron bar every six hours and was difficult to dispose of. And engineers had yet to devise a safe method of handling the volatile coal dust when the plant commenced operations.

As one citizen of the day observed, "You can hide the fire, but what are you going to do with the smoke?" Charles Dickens, who at twelve had been forced to work in a blacking warehouse after his father was imprisoned for debt, pulled no punches. In his embittered novel *Hard Times*, published in 1854, the writer created an archetypal industrial community invaded by tall chimneys: "It was a town of red brick, or of brick that would have been red if the smoke and ashes had allowed it; but as matters stood it was a town of unnatural red and black like the painted face of a savage. It was a town of machinery and tall chimneys, out of which interminable serpents of smoke trailed themselves for ever and ever, and never got uncoiled. It has a black canal in it, and a river that ran purple with ill-smelling dye, and vast piles of buildings full of windows where there was a rattling and a trembling all day long, and where the piston of the steam-engine worked monotonously up and down like the head of an elephant in a state of melancholy madness." The town's beleaguered citizens had themselves been transformed into subhuman machines in eerie anticipation of George Orwell's *Nineteen Eighty-four*: "It [was] . . . inhabited by people equally like one another, who all went in and out at the same hours, with the same sound upon the same pavements, to do the same work, and to whom every day was the same as yesterday and tomorrow, and every year the counterpart of the last and the next."

In a lecture delivered at Bradford's school of design in March of 1859, the English critic and social theorist John Ruskin evoked the image of the chimney to characterize what he feared would become of his country in fifty years: "that from shore to shore the whole of the island is to be set as thick with chimneys as the masts stand in the docks of Liverpool: that . . . you shall travel either over the roofs of your mills, on viaducts; or under their floors, in tunnels; that, the smoke having rendered the light of the sun unserviceable, you work always by the light of your own gas: that no acre of English ground shall be without its shaft and its engine."

Nor were the novelist and the aesthete without their allies among the working class. Some thought the chimneys "supremely ugly," while others created images of them that found their way into the emerging industrial folklore. One story transformed them into the long, slender necks of subterranean monsters that erupted through the soil to exhale their fetid and unnatural breath, much like the dragons of Arthurian times.

Although no one rhapsodized over tall chimneys with the eloquence Stevenson displayed for cathedrals, there were those who contemplated the new structures with a sense of awe. After taking the measure of the giant stack at St. Rollox, one secular pilgrim noted that while it is difficult to be poetic about smoke, and soda, and sulfur, and salt, and soap, "the chimney of this vast chemical establishment is something beyond the prose of ordinary street-walking mortals." In France the industrial chimney became the subject of art at the hands of several painters, including Edgar Degas, whose pastel *Landscape with Smokestacks* is considered a minor masterpiece.

Architects too were gradually drawn to the tall stacks by the challenge of combining aesthetics with altitude, not unlike the cathedral builders of old. For this, many turned to the style known as Italianate, which was first employed in the watch or bell towers of medieval Italy.

Rendered in either brick or stone, the industrial campanile was a four-sided structure crowned by a turret and narrow arched windows behind which no watchman's shadow ever appeared. One of the earliest, the Bank Works of London (1846), had spiral stairs that circled the flue, eventually depositing the persistent climber on a balcony at the top. Brick and tile manufacturers were especially attracted to the style because it enabled them to advertise their wares by incorporating glazed terra-cotta and other materials into the shaft and the ornamental capping. Within a decade or two England was replete with copies of architectural masterpieces à la Giotto, Pisano, Lamberti, Sansovino, and others.

Some architects looked considerably farther back into history for inspiration. Whatever else may be said about Napoleon's Egyptian campaign, it launched a European craze for all things associated with the land of the pharaohs, from mummies to crocodiles, furniture to makeup. The drawing boards of England's architectural firms were soon covered with modified sketches of the giant obelisks dedicated to the Sun god Aton. Among the most frequently copied was one dating from the reign of Ramses II, now overlooking the Place de la Concorde in Paris. Equally compelling as architectural motifs were Cleopatra's Needles, the popular name for two nearly 70-foot shafts of red granite erected at Heliopolis by Thutmose III, and later transported to Alexandria by order of the Roman emperor Augustus.

This slender, four-sided tapering form, terminating in a pointed or pyramidal top, provided camouflage for an ungainly smokestack in about as rational a manner as could be conceived. It was particularly attractive to newly rich industrialists, who were desperate to establish their cultural credentials as patrons of the arts. In an ironic twist, one of Cleopatra's Needles was sent to England in 1878, a gift of the profligate Egyptian ruler Ismāʿīl Pasha. It now stands forlornly on the Thames embankment. The other arrived in New York City two years later and was installed in Central Park. Because of air pollution, the hieroglyphic inscriptions of Thutmose III and Ramses II covering its sides have suffered greater erosion since the stone shaft reached the Western world than in the previous thirty-four centuries.

The end of romantic Victorian chimney construction coincided with the end of the century, when a copy of Giotto's Florentine bell tower was built in Leeds in 1899. Such structures, together with the towering behemoths of previous decades, had fallen victim to the very technological progress they had ushered in. Engineers had gradually come to the understanding that by designing more efficient furnaces, boilers, and flues, a chimney of 200 feet or less could do the job of one nearly twice as tall, and do it even better.

These changes were paralleled by a rapid decline in the use of traditional brick and stone. In 1889 a 280-foot wrought-iron shaft was erected at Le Creusot in France, its eight rings of curved plates riveted together in a matter of weeks. Like thousands of similar chimneys, most of which were shorter by 100 feet or more, the Le Creusot giant was prefabricated at a foundry and then shipped to the construction site for assembly by the numbers.

While the largest of the metal structures were belled at the bottom like their brick counterparts, it was even cheaper to employ a straight steel tube supported by a set of guy wires attached two-thirds of the way up the shaft. In either case, the savings, in both material costs and labor, were considerable. A metal chimney occupied less ground space than a brick one, required a less substantial foundation because of its relative lightness, and offered less resistance to the wind than a brick shaft of equal flue area. Also appealing was the fact that it could easily be disassembled and moved to another site on short notice.

What was being forged in iron was also being cast in concrete. Far surpassing the British in this area, the U.S. firm of Weber Company erected over 1,000 cement chimneys, all prefabricated, between 1903 and 1910. The prospective buyer had only to peruse a catalog supplied by the manufacturer before selecting the design best suited to his purposes, much as the housebound shopper agonized over the latest edition of Sears and Roebuck's "wish book." As William Wallace Christie, the mechanical engineer and author of *Chimney Design and Theory* (1889), observed, "The most prominent feature of the world at large, of every steam or power plant, and that by which the manufacturing character of a village or city is most easily distinguished, is the chimney." Wreathed by its dark aura of smoke, not even the chiseled decrees of the pharaohs could stand.

6

VULCAN'S ANVIL

~

An age employed in edging steel
Can no poetic raptures feel . . .
No shaded stream, no quiet grove
Can this fantastic century move.

—Philip Freneau, *On the Emigration to America
and Peopling the Western Country*

In 1787 an Italian visitor to England made his way to the Severn Gorge at twilight and paused to gaze down at the river 330 feet below. He then undertook the slow descent to the industrial village of Coalbrookdale, the seeming incarnation of Vulcan's anvil and perfect symbol of the new energy-hungry age. "A dense column of smoke arose from the earth; volumes of steam were ejected from the engines; a blacker cloud issued from a tower in which was a forge; and smoke arose from a mountain of burning coals which burst into turbid flame. In the midst of this gloom I . . . passed under a bridge constructed entirely of iron. It appeared as a gate of mystery, and night, already falling, added to the impressiveness of the scene."

According to archaeologists, the ability to make iron dates back at least 4,000 years. The first individuals to do so learned to heat a small amount of ore in a charcoal fire whose temperature had to reach 2,800°F before melting could occur. This slow, tedious process yielded such small quantities of the metal that it was first shaped into jewelry for royalty, who valued it more highly than gold.

Over time, it was observed that the fire burned brighter and more intensely on a windy day, speeding up the magical process of transformation. Bellows were later fashioned from animal skins to create an artificial wind, or "blast," as iron makers would one day call it, and were worked by men who took turns pumping them with their feet. The next step involved the making of small clay furnaces into which air was injected through strategically placed holes. As technology improved and production rose manyfold over the ensuing millennia, iron gradually displaced bronze as bronze had displaced stone. Utilization of iron for weapons put arms in the hands of the masses for the first time and set off a series of large-scale movements of peoples that did not end for 2,000 years, changing the face of Europe and Asia forever.

The monks of Buildwas Abbey were operating a small ironworks in the Severn Gorge when Henry VIII dissolved the religious orders in 1536. Their place was taken by craftsmen, who employed new techniques for utilizing molten iron. They made molds in the ground, lined them with sand, and then filled them with the liquid metal, a process known as casting. Once the iron had solidified, it was removed from the mold, burnished, and passed on to an itinerant salesman, or iron-monger, who sold the andirons, firebacks, and pots door to door.

To meet the increasing demand for cast iron, new types of furnaces were built with stone or brick walls lined with special clays able to withstand high temperatures. Human-powered bellows were replaced by larger ones driven by a waterwheel, causing the fire to burn even more intensely. The impurities in the ore, which had a significant effect on the quality of the finished product, were reduced by the introduction of limestone. Once the limestone melted, certain unwanted substances, including phosphorus, sulfur, manganese, sand, and dirt, combined with the liquid to form slag. Because slag is lighter than iron, it floated to the surface and was skimmed off. A perceptive observer could usually tell the age and the productivity of an ironworks by the size of its slag heap.

When the molten iron was ready, the furnace was opened at the bottom, permitting the glowing, white-hot liquid to flow into sand beds formed into molds. These were located side by side at right angles to the main channel and reminded the workmen of piglets being suckled by a sow, hence the term "pig iron." The process of making pig or cast iron was called reduction by some, smelting by others.

True to its name, the land surrounding Coalbrookdale contained large deposits of coal, as well as the iron ore used in smelting. Yet this cheap and ready fuel source could not be exploited in the blast furnace. As we have seen, coal has a high sulfur content, and sulfur combines with the iron during smelting. A blacksmith who attempts to work coal-fired iron will literally see the brittle metal crumble beneath his hammer blows.

The only satisfactory heat source was charcoal, which had to undergo a "refining" process of its own, requiring a minimum of five days for the heaps of smoldering wood to blacken and dry, leaving a residue of pure carbon. This process not only was time-consuming but became ever more costly as ironmasters vied with householders for the coveted fuel, the timber for which was being felled at an alarming rate.

Here matters stood when, in 1708, a thirty-two-year-old Quaker moved his family to the bustling Shropshire village. Abraham Darby had come to Coalbrookdale from Bristol, where, until recently, he had been partners in a foundry for casting "bellied" iron cooking pots in sand, a marked improvement over the loam and clay molds then in use. Darby patented the process, but his partners did not share the young inventor's enthusiasm. Deciding to go it alone, Darby leased a small blast furnace on the Severn, where he dreamed of turning out iron pots by the thousands, the profits from which would be entirely his own.

Darby began smelting with charcoal and was immediately faced with the problem of its high cost, a daunting barrier to profitable operations. He had been in Coalbrookdale no more than a year when he began making entries in his accounts for the purchase of "Coles" and "charking coles." While it is impossible to tell if these are references to plain coal, charcoal, or both, there is little question that Darby was in the throes of experimentation.

At some point Darby had begun heaping coal in the open air and setting it alight. The resulting heat drove off the natural gases, leaving a hard, porous, massive residue of nearly pure carbon called coke. Darby was quick to recognize the similarity between coke and charcoal, which were interchangeable insofar as smelting pig iron was concerned. By stoking his furnace with the new fuel, Darby had taken the first, and most difficult, step in freeing the iron industry from the tyranny of charcoal. The Shropshire Quaker was able to sell his pots, kettles, and other metal products at prices within the reach of the average citizen, and for this, history would remember "Darby of the Dale" as a great benefactor to the English people.

Darby neither patented his process nor, apparently, did he keep it secret, yet only six other coke furnaces were built in all of Britain during the next fifty years. Ironmasters who tried to duplicate his method usually failed for reasons that neither they nor Darby fully understood. The most important reason was that Darby used a coal with a very low sulfur content, which was not true of that burned by most other smelters, causing their inferior utensils to crack during heating in the family hearth.

The next major problem Darby encountered—and was unable to resolve—concerned the production of malleable bar iron sold to forgemen for making into edge tools, plowshares, wagon rims, hoes, hinges, ornamental work, and other products. For this production, coke smelting proved unsatisfactory because the carbon content of the iron rendered it too brittle for shaping on anvil and forge. After many experiments, none of which were successful, Darby was still purchasing charcoal for use in reprocessing iron for blacksmiths. So long as other ironmasters had to switch to charcoal from coke to convert pig into bar iron, they were reluctant to embrace the new technology, remaining faithful to the old ways, especially when they were located near a ready source of timber. The few who were successful in smelting cast iron with coke usually did so on the sly to guard their financial interests against the suspicious forgemen.

Darby's eldest son, Abraham II, was barely six when his father died at age forty. He entered the family business as a youth of nineteen, seeking to enlarge the market for coke-fired iron. At the time, the nail industry located in the Midlands was the largest consumer of the metal, but its owners refused to purchase any iron not made with charcoal.

Around 1749, Darby, who was now in his late thirties and had greatly expanded the Coalbrookdale Works, began to experiment with casting malleable iron. His highly skilled workmen put in twelve-hour shifts, from six till six, seven days a week, keeping the furnace in continuous blast while Darby paced the overhead bridge like a sea captain transfixed by the Maelstrom. The story is told that for six days and nights he refused to leave the walkway, ignoring the pleas of his family and men to go home and rest. Then came the flow of iron he had long been searching for, after which the spent owner collapsed and had to be carried to his new house in the dale.

Darby had at last succeeded in producing a quality of malleable iron acceptable to both the nailers and the forgemen—and he had done it by burning coke. The precise technique he employed was never revealed, although it is suspected that his secret lay in the selection of iron ore with a low phosphorus content. What we do know is that the two Abrahams had so improved the art of smelting with coke that after 1760, the quality of their iron was favored by housewives and manufacturers alike. In the next half century charcoal furnaces would all but disappear in England, and by about 1800 the quantity of the world's coke production had increased from virtually nothing to more than 100 million tons a year.

If there was ever a place representative of the industrial revolution in microcosm it was Coalbrookdale. Newcomen engines, their boilers cast by the score in the throbbing ironworks, drove its huge flywheels and giant bellows while pumping water from the local mines. By the 1780s, when Abraham Darby III assumed control, the company had begun to replace these aging monsters with the more efficient and powerful Boulton and Watt engine.

Of all his many achievements, Abraham III took the greatest pride in the design, casting, and construction of the Iron Bridge, or "gate of mystery" as the entranced visitor from Italy called it. Made up of five separate arching ribs, each of which was cast in two pieces and lowered into position to form a single graceful span, the structure contained not a single rivet or screw. The men of Coalbrookdale designed cast-iron joints like those employed by a master carpenter hewing a piece of fine furniture. A roadway of thick iron plates was then laid across the elegant superstructure, edged by ornamental iron fencing. So expert was the engineering and craftsmanship that to this day the Severn's deep waters slip quietly beneath the span as they have for two centuries and more.

~

Shropshire, like the rest of England and neighboring Wales, was webbed by canals built to provide an inexpensive means for transporting coal, increasing amounts of which were being carried from mines via rail. The open wooden cars used in hauling were fit with flanged iron wheels, which kept them on a set of parallel tracks (called wagonways) that made them self-steering. Until the nineteenth century, they were pulled by humans or horses.

It was not by accident that Richard Trevithick, a young Cornish engineer, found his way to Coalbrookdale in 1802. The Darbys had pioneered the casting of iron wheels for use on their own ore-carrying wagons; then, in 1767, they laid a strip of iron on the tracks to reduce the wear caused by the metal wheels. Within twenty years the composite rails were being replaced by tracks made completely of iron.

Trevithick was bent on building the high-pressure steam engine that James Watt refused to manufacture because he believed that it would surely explode. Although Trevithick was aware of the danger, he realized that such an engine would be much lighter than anything Boulton and Watt had made, thus opening new commercial possibilities. Napoleon's rampage across Europe had caused the price of animal feed to rocket. Why not replace live animals with an "iron horse," one that mine owners could nourish with their ample supplies of coal? Besides, Trevithick had started conducting his own experiments on high-pressure steam five years before he arrived in Coalbrookdale and had actually built a locomotive, which he tested successfully near his home at Camborne.

With parts supplied by the Coalbrookdale Works for a stiff £246, Trevithick built his dream engine. After tests showed that it could be operated safely at a pressure of 145 pounds per square inch, he placed it on a car with wheels and dubbed it the *New Castle*. Its horizontal boiler was surrounded by huge iron cogs and gears reminiscent of the clockwork universe that sparked the scientific revolution. Hoping to generate financial support, the inventor built another steam engine cleverly named *Catch Me Who Can*, which almost anyone could do, for it moved little faster than a walk. He displayed it in London, in 1808, behind an enclosed circular fence. Passengers could ride the circuit for a shilling apiece in an attached road coach refitted with special wheels. While the novel machine sparked public interest, it produced no nibbles from potential investors, forcing a crestfallen Trevithick to turn his attention to more profitable aspects of engineering.

At Killingworth mine, near Newcastle, a young engine-wright named George Stephenson had also undertaken a study of Watt's steam engine. His conclusions were sufficiently promising to gain the financial backing to construct a locomotive named the *Blucher*, after the delusional but brilliant Prussian field marshal who had been giving Napoleon fits for years. Following a successful trial run in 1814, Stephenson secured a patent for the "steam blast," by which hot exhaust was directed up the chimney, pulling air after it and increasing the draft. He designed and built more powerful locomotives in the following years and supervised the construction of an eight-mile railway for the Hetton colliery, which opened in 1822.

The true test came with the construction of the Liverpool and Manchester Railway over the unheard-of distance of thirty miles. Stephenson was hired as consulting engineer and faced a number of obstacles that taxed his ingenuity to the limit. He decided to elevate the roadbed above the level of the Lancashire plain to improve drainage and reduce grades; a causeway was pushed across Chat Moss swamp at a cost of more than £40,000 per mile; a long tunnel from the Liverpool docks dropped the line below the level of the city. Parliamentary intervention was required when surveyors and construction crews were threatened with bodily harm by worried landowners and their equally anxious tenants, grist for George Eliot's *Middlemarch*.

All this was done without any assurance that the venture would succeed, a measure of the optimism born of the new technology. What was needed was a reliable locomotive—one, according to rules of the competition announced by the railway's directors, that could achieve a mean speed of ten miles per hour with a steam pressure not exceeding fifty pounds per square inch. The winner of the Rainhill trials, scheduled to begin on October 6, 1829, would receive £500 and the honor of having his engine inaugurate the first fully evolved railway service.

Stephenson's locomotive, one of only four entered in the trials, was built by his son, Robert, in their Newcastle works. Dubbed the *Rocket*, it was anything but streamlined and looked more like an oversized wheelchair than a self-propelled vehicle. The boiler, which measured six feet in length and forty inches in diameter, was mounted on two sets of wheels, one twice as large as the other. It sported a tall chimney and was driven by two large cylinders placed obliquely to the axis. Its greatest advantage was that it weighed only four and a quarter tons, much less than most prototypes constructed to date.

The other competitors—*Novelty, Sanspariel,* and *Perseverence*—were not ready to be timed on opening day, leaving the field to the *Rocket*. It ran a steady twelve miles in fifty-three minutes, close to expectations that the locomotive was a "sixteen-mile-an-hour wonder." None of the others mounted a serious challenge, and the Stephensons were hailed by the crowd of enthusiastic onlookers when they stepped up to claim their

George Stephenson's Rocket
(Used by permission of Culver Pictures)

prize. According to the *Scotsman*, "The experiments at Liverpool have established principles which will give a greater im-
pulse to civilization than it has ever received from any single
cause since the press first opened the gates of knowledge to the
human species at large."

Within a decade of the *Rocket*'s triumph, more than 1,300
miles of track were laid in Britain, most of it radiating from
greater London. Continental Europe was quick to follow, with
Stephenson supplying both the plans and the technology to
construct the world's first international line between Liège
and Cologne. By building an extensive system of its own,
Prussia ultimately forced a unification of the German states.
In like fashion the Kingdom of Piedmont, through its rail-
ways, brought pressure on the Italian states to form a unified
country. Shouldering their way across the prairie, spanning
great rivers, and bulling through mountains, some of which
were as yet unnamed, the Central Pacific and Union Pacific
railroads met at Promontory, Utah, in 1869, where the driving
of the symbolic golden spike reverberated across the North
American continent. A "great and shining road," its awed ad-
mirers called it, stretching from sea to shining sea.

7

THE PHANTOM
OF THE OPEN HEARTH

~

Double, double, toil and trouble;
Fire burn and cauldron bubble.

—William Shakespeare, *Macbeth*

It was Henry Cole, a well-connected London civil servant and tireless champion of free trade, who conceived of the ambitious idea. Exhibitors from all the civilized nations would be invited to take part in an international exposition—a world's fair—devoted to the fruits of manufacturing and technology. Cole approached Prince Albert, consort to Queen Victoria, who immediately grasped the financial implications of such a venture, for Britain had far more to gain than any of its competitors. When the queen was informed, she agreed to have her name placed at the top of the subscribers' list. The Royal Society of Arts, of which the prince was president, added its approval, and the necessary funds were soon raised.

The first thing required was an exhibition hall, which was to be erected on a roughly twenty-acre site in Hyde Park. Time was of the essence; the Great Exhibition of the Works of Industry of All Nations was scheduled to open in the spring of 1851, leaving little more than a year to come up with a satisfactory design and see it through construction.

A building committee was hastily formed and announced an open architectural competition. It envisioned a temporary structure covering sixteen acres and was soon deluged by 245 entries, none of which were deemed acceptable. In desperation, the committee proceeded to cannibalize the drawings, taking what its members considered the best features of each to create their own blueprint. A sketch of the resulting camel, complete with a domed hump 200 feet high, was published and immediately attacked by the *Times* as an affront to good taste.

Enter Joseph Paxton, the son of a Bedfordshire farmer, who had risen to become superintendent of the gardens at Chatsworth, the ancestral estate of the duke of Devonshire. Paxton had already displayed a special talent for structural design, having built a giant greenhouse enclosing an acre of ground. The Great Conservatory, as it was called, was praised by visitors as a "modern marvel." Inside the protective glass was *Victoria regia*, a giant water lily whose seeds had been brought back from tropical British Guiana. The royal gardeners at Kew had been unable to make it grow, but it thrived under Paxton's care, producing huge leaves and beautiful flowers, one of whose buds was presented to the queen for whom Paxton had named it.

One day, acting on a whim, Paxton picked up his young daughter and gently set her down on one of the huge leaves. Instead of collapsing and plunging the child into the shallow water, it remained rigid, which Paxton attributed to the geometric pattern of thick ribs and crossribs on its underside. With pad and pencil, he sketched the outline of a glass house to be built especially for the lily, based on the design of its own leaf. Once executed, the crystal structure became the perfect environment for the massive plant.

Paxton had not entered the competition and was surprised when he learned that no acceptable design had been submitted. With only two weeks to go before an official announcement, he sketched a plan of his own during a business meeting. After completing a more detailed rendering a week later, he persuaded the committee to accept his belated entry.

The fact that Paxton had friends in high places was sufficient to tip the balance. After first pooh-poohing the new design, the committee thought it more promising after a second look. Among the proposed structure's advantages were its extreme simplicity, the speed at which it could be assembled, and the absence of internal walls. A sketch of it was prematurely published in the *Illustrated London News*, sealing the camel's fate. The committee announced that it was abandoning its own cumbersome plan in favor of Paxton's.

"Ferro-vitreous" was Paxton's medium—iron and glass. The totals are nearly as astounding today as they were in 1851: 3,300 hollow cast-iron columns; 2,224 principal girders; 24 miles of main gutter; 205 miles of wood sash bar to hold the transparent roof panels in place. Over and along the sides of this gigantic skeleton, which covered 772,824 square feet—or about 19 acres—were 900,000 square feet of sheet glass, equal to a third of England's total production in 1840. A name, the perfect name, for this fairy-tale structure came suddenly to mind: the Crystal Palace.

It would become the most influential building of the nineteenth century, indeed one of the most influential structures ever erected, taking its place with a handful of others such as the Parthenon, Hagia Sophia, and Abbé Suger's Saint-Denis. Possessed of no truly monumental feature, it was itself monumental in a purely quantitative and aesthetic sense. The main body of the building was 1,848 feet long and 408 feet wide and had an addition on the north side measuring 936 by 48 feet. Its longitudinal central aisle, known as the "main avenue," was 72 feet wide by 66 feet high, in the middle of which was the arched transept reaching a height of 108 feet. The transept had been added to keep from having to cut down a cluster of 90-foot elms, whose cause was championed by Colonel Charles Sibthrop, an avid, if idiosyncratic, conservationist.

The Crystal Palace was entirely prefabricated in standardized units. The panes in the "ridge and valley" roof inspired by *Victoria regia* were exactly forty-nine inches long. Each was made by blowing liquid glass onto a hollow cylinder, cutting it longitudinally, and allowing it to open up and flatten in a kiln. To simplify construction, workers riding in specially designed wheeled trolleys set the panes in the sloped wooden "Paxton gutters" installed in valleys to carry rain and condensation to the hollow iron columns that doubled as drainpipes.

The thin but strong cast-iron columns were largely a product of the new railway age, which called for the construction of hundreds of iron bridges. Besides holding up the glass walls and roof, they supported the elevated walkways that led to an additional 200,000 square feet of promenade and exhibit space.

With the ability of workmen to raise dozens of columns and install thousands of square feet of glass each day, the Crystal Palace was completed in a remarkably brief seventeen weeks, becoming the world's first large freestanding iron-frame building, the essence of the "skin and bones" style later envisioned by architect Ludwig Mies van der Rohe and embodied in his Seagram Building. Indeed, a contemporary of Paxton's suggested that the modular units be rearranged to form a 1,000-foot tower, a vertical Crystal Palace that would have been feasible but for the fact that cast-iron columns could not have carried the weight, requiring the use of steel beams, which were yet to be invented. There was also the question of elevators.

Even as the work neared completion, the naysayers would not be silenced. Surely the first spring storm, especially if accompanied by hail, would reduce the glass box to a mountain of shards. Failing that, the summer heat and humidity would make it unbearable. Neither prediction came to pass. The Crystal Palace mastered the elements, remaining cool and dry thanks to Paxton's gutters and swaths of canvas suspended above the roof.

The Great Exhibition was officially opened on May 1, 1851, with much pomp and circumstance by a beaming Queen Victoria. More than 6 million visitors would attend before the closing ceremony in mid-October, and this despite the fact that the Crystal Palace was closed on Sundays. They were greeted by 14,000 exhibitors, over half of whom were British, and an estimated eight miles of display tables. After much grousing about Anglo-Saxon effrontery and threats of a boycott, the French gave in and sent 1,760 exhibits. The United States, still a fledgling nation in the eyes of most Europeans, supplied another 560, including false teeth, artificial limbs, sulkies, a McCormick's reaper, the Colt revolver, chewing tobacco, and Goodyear India rubber products.

Not only did the British learn about the best arts, crafts, and manufactures of other peoples, visitors from the Conti-

The Crystal Palace
(Used by permission of Culver Pictures)

nent and the rest of the world were made acutely aware of the superiority of British goods, machinery, and production techniques. Orders poured in, while the Great Exhibition's sponsors reaped a profit of $750,000. So taken was the queen by all of this that she returned to the Crystal Palace some fifty times before its closing. On the final day, Victoria wrote wistfully in her journal, "To think that this great and bright time is past, like a dream, after all its success and triumph."

The real dream was the ethereal greenhouse, almost as invisible as the atmosphere itself. It was soon to be dismantled and reassembled atop 200 wooded acres on Sydenham Hill, south of the city, where it would continue to dazzle until it was destroyed by fire in 1939.

Little did the irreverent editors of *Punch* know how prophetic they were when they quipped, "We shall be disappointed if the next generation of London children are not brought up like cucumbers under a glass." Paxton, who was knighted and made a member of Parliament for his creativity, had more than met the goal of his profession by fusing transparency, immateriality, and indefinite boundaries to mask all evidence that one was strolling under an invisible and fragile roof. The Crystal Palace as environmental metaphor was fully a century ahead of its time.

~

The Great Exhibition neatly separated the nineteenth century into two halves, indeed into two very different historical times—agrarian and industrial. The future was well represented in the form of a diminutive, feisty, cigar-smoking Londoner named Sir Isambard Kingdom Brunel, who trod the glass gallery with a swagger befitting his weighty appellation. Not only was Brunel a zealous promoter of the fair, he was a member of the building committee, as well as chairman and reporter of the section on civil engineering. Brunel was particularly interested in any exhibit that had to do with metal and its manufacture, especially iron. He paused to pay his respects and talk shop with the latest generation of Darbys, whose display of wrought iron was one of the exhibition's most impressive.

If Brunel had had his way, and he usually did, the streets, the sidewalks, the very buildings of London would have been constructed of iron; his addiction to metal made him one of the most important, if largely forgotten, figures of the nineteenth century. He had come of age under the guidance of his father, the civil engineer Sir Marc Isambard Brunel, who, in 1824, with the support of a company formed by the duke of Wellington, contracted to build a tunnel under the Thames from Rotherhithe to Wrapping. Although Isambard was only nineteen at the time, he was appointed resident engineer when work began a year later.

The plans called for a passage consisting of large double archways, so that traffic could flow as easily under the river as upon it. Sir Marc patented the design for a shield capable of covering the total area to be excavated. It was divided into twelve separate compartments made up of thirty-six smaller cells in which the miners labored independently of one another to guard against a catastrophe. The gigantic apparatus was advanced inch by inch with screw power as the excavation proceeded.

That the work was dangerous is a gross understatement. The Thames breached the shield at least five times in the seventeen years it took to complete the project, panicking the workmen and claiming several lives. At one point construction was stopped, and the tunnel was bricked up for seven years. The cumulative strain on the elder Brunel finally triggered a stroke followed by partial paralysis, but he recovered sufficiently to take part in the opening ceremony in March of 1843, the Iron Duke at his side.

The son drove himself just as hard. During an anxious period in September of 1826, Isambard stayed on the job for ninety-six consecutive hours, taking catnaps inside the tunnel. In another instance, when the river suddenly began pouring in, trapping a miner, he lowered himself into the shaft, which was already half filled with water, and brought the man to the surface unharmed. In 1828 he had a second brush with death and was seriously injured in another irruption.

His reputation for tenacity now established, Isambard was appointed engineer to the Great Western Railway in 1833. He proceeded to construct broad-gauge tracks (seven feet apart as opposed to the standard four and a half feet), arguing that wide-set wheels would let travelers read their newspapers while sipping tea undisturbed at forty-five miles per hour or more. This provoked the infamous "Battle of the Gauges," which was based as much on ego as sound engineering principles. It was a fight that Brunel ultimately lost, but not before he built the world's fastest railroads, the world's longest tunnel, and the world's most daring bridges. Brunel was simply the greatest practical builder and engineer in British industrial history, well deserving of his Victorian nickname, the "Little Giant."

As a man who thrived on challenges, Brunel saw no good reason to stop building just because the land gave out. Following in the footsteps of his father, a pioneer in the application of steam power to navigation, Isambard built the three most advanced oceangoing steamships of his time, each the world's largest at its launching. The *Great Western*, a double paddle wheeler, was the first to come off the ways. At 236 feet in length and 35 feet in breadth, with a displacement of 2,300 tons, this ship was almost 30 feet longer than the next-largest steam vessel when it made its first transatlantic crossing in 1838. The voyage was completed in the unprecedented time of fifteen days, thus inaugurating regular steamship service between England and the United States, otherwise known as "the Atlantic Ferry."

In the meantime, Brunel had been commissioned by the Admiralty to conduct experiments using the screw propeller. The results were so promising that he incorporated screw propulsion into the design of his second behemoth, the *Great Britain*. Equally radical was his decision to abandon the traditional wooden-hull design in favor of iron, which critics argued would make the ship so heavy that it would exhaust its coal supply and become stranded in midocean. Brunel understood what today's supertanker contractors know: The longer and larger a well-shaped hull is, the more efficiently it can be propelled through the water. The *Great Britain* took to sea in 1845, completing its maiden voyage from Liverpool to New York without incident—and with fuel to spare.

By the mid-1850s, Brunel was passionately immersed in his two greatest construction projects, both involving iron and steam. The directors of the Cornwall Railway, hoping to bridge the river Tamar at Saltash, contracted for Brunel's services. The result was two spans, each 455 feet long, and a central pier supported by girders anchored to rock a dizzying 80 feet below.

At the same time, Brunel was engineering the construction of his third and last steamship—the gargantuan *Great Eastern*—the first vessel to boast a double iron hull and propulsion by twin screws, complemented by two 58-foot side paddles. At nearly 700 feet, the ship was double the length and triple the tonnage of the largest vessel then afloat, while its massive hull, consisting of thousands of 700-pound riveted iron plates, was the heaviest object humankind had ever attempted to move. When fully laden, its 22,000 tons would outweigh all 197 ships that had been sent out to confront the Spanish Armada. The five-foot, four-inch Brunel was photographed with top hat and cigar standing next to the vertical anchor chain, only five links of which were needed to match the height of the proud engineer.

There were skeptics as always. An unimpressed Herman Melville, who remained steadfast to the tall ships, wrote sarcastically in his journal, "Vast toy. No substance." To another observer, the "Eighth Wonder of the World" seemed to "weigh upon the mind as a kind of iron nightmare."

Brunel's dream was to have his great ship become a luxury passenger vessel, plying the Atlantic to a backdrop of Viennese waltzes, clinking champagne glasses, and haute cuisine. When he limped painfully aboard his "great babe" for the last time, the ship was still a few days away from embarking on its shakedown cruise. Brunel collapsed minutes later, the victim of a stroke like his father, and was carried home. Four days after his seizure, a faulty steam valve caused a funnel casting to explode on the ship's first foray into the English Channel, resulting in the death of several people. His spirit broken, Brunel succumbed within a week.

The *Great Eastern* never lived up to its promise as a luxury liner; the number of people with sufficient resources to book travel on the vessel were simply too few, a reminder that the industrial revolution did not create millionaires overnight. The ship's greatest service was as the vessel that laid the first Atlantic cable in the summer of 1866. To make this possible, dockyard workers tore out staterooms and walnut-and-velvet saloons, replacing them with circular tanks large enough to hold all 2,500 miles (7,000 tons) of cable needed for the western crossing to Canada. Following an uneventful two-week passage, the triumphant ship sailed into Newfoundland's Trinity Bay, sparking celebrations on both sides of the Atlantic. President Andrew Johnson and Queen Victoria exchanged the first telegrams free of charge; everyone else had to pay four shillings a word. The *New York Sun* declared it "the grandest [work] which has been accomplished in our age."

~

No one predicted that the annual meeting of the British Association for the Advancement of Science would produce a bombshell in August of 1856, let alone that it would be delivered by a man whose formal education had ended with grammar school. Just who was this Henry Bessemer, anyway, and exactly how had he been able to master "The Manufacture of Malleable Iron and Steel Without Fuel," as he so boldly proclaimed in the title of his paper? Still, he had been granted a government patent the year before, which meant that there must be something to it.

If the iron manufacturers wanted to find out, they would have to pay—and pay dearly—which most were surprisingly eager to do. Within a month of the meeting, Bessemer collected a staggering £27,000 for licenses to use his invention. Within another month he was greeted by a chorus of derision after hastily conducted trials failed to produce the promised result. The outraged licensees were convinced that they had been cheated. Bessemer was a charlatan! Bessemer was a fraud! Bessemer must pay up or go to jail!

In point of fact, forty-three-year-old Henry Bessemer was already a wealthy mechanical genius and inventor on his way to receiving his hundredth patent. In 1854 England and France were allied with Turkey against Russia in the Crimean War. Bessemer, who had spent a good deal of time on the Continent and was fluent in French thanks to his Huguenot ancestry, attended a party whose guest list included Prince Napoleon, cousin to the emperor Napoleon III. The prince was introduced to Bessemer as the inventor of a new method of firing a projectile and was intrigued when the Englishman reached into his pocket and pulled out a toy shell, carved from mahogany. Bessemer explained that an elongated projectile could be hurled much farther than a round one, but it must rotate if it is to achieve accuracy. Experiments with a small cast-iron cannon convinced him that his idea was feasible, yet it had been dismissed as nonsense by the generals of the British War Department.

Bessemer was soon granted an audience with the emperor himself, who gave the inventor carte blanche to conduct further experiments at the testing grounds near Vincennes. Bessemer arrived on a cold December day, his new projectiles in hand. Their firing more than met his expectations, but he was brought up short by the military officer in charge of the experiment. It was one thing to employ a small shell, but would it be safe to fire a thirty-pounder from a twelve-pound cast-iron gun? "This simple observation," Bessemer wrote in his autobiography many years later, "was the spark which kindled one of the greatest industrial revolutions that the present century has to record." Three short weeks later, he applied for the patent that gave birth to the age of steel.

Steel was being manufactured well before the bearded and

Bessemer's Converter
(Used by permission of Culver Pictures)

long-faced Henry Bessemer entered the lists, but it was a laborious undertaking whose final product sold for fifty to sixty pounds a ton, a prohibitive figure by nineteenth-century standards. The process began when pig iron was heated to the melting point in a puddling furnace. Once it was molten, workers had to continually stir, or puddle, the iron to remove the excess carbon while raking off the slag. This took well over a week and consumed large amounts of fuel. After the excess carbon was removed, the white-hot metal could be rolled into large 100- to 200-pound lumps called blooms. These were then removed to a steam-driven forge for rolling and hammering into the desired shape. Because it was extremely hard and would take an edge, the finished steel was used almost exclusively for making the finest cutting tools.

In attempting to produce a cast-iron cannon strong enough to fire a powerful shell, Bessemer discovered that the carbon removed during puddling could be eliminated in another, and far simpler, way. He built what he termed an experimental "converter" in the shape of a giant egg, with many small holes in the bottom, and filled it with molten iron. Unaware of the risk he was taking, Bessemer subjected the red-hot brew to a powerful blast of air, hoping that the carbon it contained would burn off. The sudden injection of oxygen unleashed a "veritable volcano" of flames, sparks, and melted slag, putting an awed Bessemer in mind of the sorcerer's apprentice. When the uncontrolled eruption finally died down more than ten minutes later, the anxious inventor poured the liquid metal into a ladle and found it to be wholly decarbonized and malleable. He had converted pig iron directly into steel in a matter of minutes!

Bessemer's triumph was short-lived. Virtually every attempt on the part of his licensees to duplicate his results ended in failure. Smelling blood, the press weighed in, calling Bessemer a "wild enthusiast" whose so-called invention had flitted across the "metallurgical horizon," only to "vanish in total darkness." Worse still, he faced financial ruin in the courts, forcing him to cancel his licenses.

One day, as Bessemer was reflecting back for the umpteenth time on what might have gone wrong, the answer struck him "like a bolt from the blue." He headed for his well-equipped laboratory, where he subjected samples of the iron employed in the failed experiments to chemical analysis. Sure enough, they all contained high levels of phosphorus, rendering them unfit for making steel. What was more, the ore from which they were smelted had been mined domestically. Bessemer, on the other hand, had used largely phosphorus-free iron imported from Sweden.

Being able to prove that he was not the fraud portrayed in the press was only half the battle. Bessemer still faced the problem of how to remove the unwanted phosphorus from domestic ore. The answer, he believed, lay in the composition of his furnace lining, which was made of fireclay. Drawing deeply on his own investments, he erected a new works in the great steel citadel of Sheffield, the very heart of enemy territory. During the year following his public disgrace, Bessemer spent thousands of pounds in the construction of new furnaces, tearing down one after another while "those most near and dear to me grieved over my obstinate persistence."

At the same time, he had shrewdly hedged his bets by importing large quantities of iron ore from Sweden. Despite the higher transportation costs, the firm of Henry Bessemer and Company Ltd. was able to sell its steel ingots at ten to fifteen pounds a ton below those produced via the old method. As one observer noted, "This was the kind of language the Sheffield steelmasters understood." They again turned to a vindicated Bessemer, who likened their zeal to obtain new licenses "to a wild pack of hungry wolves, fighting with me and with each other, for a share of what was to be made of this new discovery."

By 1859 many of the technical problems associated with the smelting of ore high in phosphorus had been resolved by others, including Robert F. Mushet, the American William Kelly, and Sidney Gilchrist Thomas, who lined his converters with limestone rather than fireclay, thus reducing the phosphorus to disposable slag. Bessemer's process, undertaken in converters that dwarfed their Lilliputian operators, had been adopted by all the steelmaking countries. At the London International Exhibition of 1862, the Bessemer steel display was one of the most acclaimed. Among its many products were locomotive boiler tube plates, rails forty feet long, large-caliber artillery pieces, the crankshaft of a 250-horsepower engine, and an ingot weighing 3,136 pounds, the 6,410th cast from the converter of the Sheffield works. And this was only the beginning. As the manufacture of steel increased exponentially, so did the royalties that flooded in from around the world at the rate of £2 per ton. Between 1866 and 1868 alone, they totaled some £1 million, sufficient to gain the resilient and tenacious Bessemer the long-coveted title of Sir Henry. The United States could do no less; six thriving manufacturing towns in Alabama, Michigan, Pennsylvania, Virginia, Wyoming, and North Carolina were named after him, with Alabama adding a county for good measure.

~

Judging from his many photographs, Andrew Carnegie could have passed as the grandfather, or even the father, of novelist Ernest Hemingway. Both possessed broad rectangular-shaped heads, the cheeks of mischievous cherubs, snow-white beards, and arresting blue eyes. However, this striking resemblance ended abruptly with their physiognomy. Hemingway was broad-chested and, in his prime, a bear of a man; at scarcely five feet, three inches, the transplanted Carnegie reminded many of a gnome, a fact that so pained the steel magnate that when he was photographed standing next to others, he preferred that it be outdoors so that he could claim the high ground.

Carnegie was born into a Dunfermline family well-read and vigorously interested in radical Scottish politics. His father, William, was a hand-loom weaver and Chartist who marched for the rights of the workingman. When the power loom arrived in Dunfermline, the family was impoverished within months. Andrew never forgot the night his father, the picture of despair, returned from a fruitless day's search for employment and said to him, "Well, Andra, I canna get nae mair work."

May of 1848 found the Carnegies bound for the United States aboard the former whaler *Wiscasset.* Their final destination was Allegheny, Pennsylvania, near Pittsburgh's famed "Golden Triangle," where the placid Monongahela meets the brawling Allegheny to form the westward-flowing Ohio.

At twelve the diminutive, towheaded Andrew became a bobbin boy in a cotton factory, laboring from six in the morning to six at night. He enrolled in night school, steeping himself in the history and literature of his adopted country, and gained his citizenship automatically when his father was naturalized in the 1850s. At fourteen he became a messenger in the Pittsburgh Telegraph office, where the pay of $2.50 a week seemed a fortune. Two years later he wrote his cousin in Scotland, "I have just got past delivering messages and have got to operating. I am to have four dollars a week now and a good prospect of getting more soon."

Thomas A. Scott, a superintendent of the Pennsylvania Railroad Company, was a frequent visitor in the telegraph office, where young Carnegie caught his eye. Scott hired him as his private secretary and personal telegrapher for thirty-five dollars a month. "I couldn't imagine," Carnegie wrote in his autobiography, "what I would ever do with so much money."

Carnegie's subsequent rise was rapid; within six years of being hired by Scott, he replaced him as superintendent of the railroad's Pittsburgh division in 1859. His greatest achievement as a railroad man was the introduction of Pullman sleeping cars. He himself acquired a one-eighth interest in the Woodruff Company, the original holder of the Pullman patents, with a bank loan of $217.50. The investment was soon to be worth more than $5,000 a year to the young capitalist, who bought stock in such industrial concerns as the Keystone Bridge Company, the Pittsburgh Locomotive Works, and the Superior Rail Mill and Blast Furnaces. By the age of thirty, Carnegie had an annual income approaching $50,000 and was on his way to becoming a multimillionaire.

Had he taken his own advice, it would never have happened. Among the papers discovered by his executors was a memorandum written in 1868: "By this time two years [from now] I can so arrange all my business as to secure at least $50,000 per annum. Beyond this never earn—make no effort to increase fortune, but spend the surplus each year for benevolent purposes."

For the time being, the voice of his father's Chartist conscience went unheeded, partly because of something Carnegie had seen during a visit to Britain in 1872. He met and became friends with Henry Bessemer, touring a steel mill in which the new converter was in operation. Resting on pivots like a giant, bulbous cannon, the twenty-foot monster tilted to take a charge of up to twenty-five tons of molten iron. When loaded, it was turned upright during the "blow," the "explosion" that occurred when air was injected through perforations in the bottom. As the air passed upward through the melted pig iron, such impurities as silicon, manganese, and carbon united with the oxygen in the form of oxides; the carbon monoxide burned off with a blue flame; the other impurities formed slag. Carnegie's wide, unbelieving eyes suddenly narrowed as the mental tumblers fell into place. Benevolence could wait.

Once back in Pittsburgh, he began divesting himself of his many financial interests to concentrate exclusively on the manufacture of steel, describing his gamble as "putting all my eggs in one basket and then watching the basket." He founded the J. Edgar Thomson Steel Works, the first plant in the United States to employ the Bessemer process. Displaying the sycophantic skill that had always served him well, he named it for the president of the Pennsylvania Railroad, a major purchaser of steel rails and a transporter of steel products.

In a further display of shrewdness, Carnegie adopted almost every financial and technological innovation that promised to reduce the cost of making steel, including the purchase of coke fields, iron-ore deposits, ships, and railroads. By 1890 the now vast Carnegie Steel Company had replaced many of its Bessemer converters with open hearth furnaces, each capable of producing as much as sixty-five tons at a time. Occasionally, when a furnace was tapped and the molten steel spewed forth, casting ghostly shadows on the girders and giant cranes, workers were greeted by the fiery visage of the Phantom of the Open Hearth, a warning perhaps that modern man was no longer within nature but a world eater, destined to consume resources exponentially.

No matter. Steel that had once sold for $100 a ton was now going for $20 and headed down to $12. By 1900 Carnegie Steel was producing more tonnage than all of Great Britain combined, reaping annual profits of $40 million, $25 million of which was Carnegie's to do with as he pleased.

Yet sooner or later there was bound to be trouble in an industry where twelve-hour shifts were the norm and the floors of the mills were so hot that the workmen had to nail protective wooden platforms to the bottoms of their shoes, a far cry from the mythical iron puddler Joe Magarac, who squirted molten steel through his fingers to form rails. Every two weeks, after putting in a double shift, they got their only day off. When they were paid, the contents of their envelopes varied according to the market price of steel. The best of their housing, as elsewhere in the industrial world, was crowded and filthy, and most died in their forties or earlier, some in violent accidents, others from disease. Carnegie and his fellow manufacturers transformed the Monongahela River into a fifteen-mile-long industrial runnel, unfit, it was said, even for microbes and bacteria; the air of the valley through which it flowed was so thick with smoke from the giant brick and concrete stacks, you could spread it on bread in the absence of butter.

While Carnegie was visiting his castle in Scotland, management decided to roll back wages because of the nationwide recession of 1892. The unions struck the Carnegie mill in Homestead, near Pittsburgh. In response Henry Clay Frick,

Carnegie's right-hand man, called in the Pinkertons, whom the unarmed strikers bested with their fists. Next came the state militia, 8,000 strong, and it was war. When the smoke cleared, sixteen were dead and hundreds more were seriously injured. Not only was the strike broken, but unionization of the steel industry would not reassert itself until 1936, during the depths of the Great Depression.

Carnegie dismissed the Homestead strike as cavalierly as he dismissed the scattered concerns about the environment. When confronted by a congressional committee on his responsibility for complying with the law, he replied, "Do you really expect men engaged in an active struggle to make a living at manufacturing to be posted about laws and their decisions, and what is applied here, there and everywhere?"

It was, after all, what Mark Twain called the Gilded Age, and, true to his word, Andrew Carnegie never once took his eyes off the clutch of golden eggs in his basket.

8

"THE DYNAMO AND THE VIRGIN"

~

He found himself lying in the Gallery of Machines at
the Great Exposition of 1900, his historical neck broken
by the sudden irruption of forces totally new.

—Henry Adams, *The Education of Henry Adams*

On the evening of Friday, September 1, 1871, the crew of the
Henry Taber, a whaler out of New Bedford, Massachusetts,
lined the deck where they witnessed a preview of their own
fate. In the near distance, the three-masted bark *Roman* could
be heard groaning like a stricken animal as it was being enveloped by giant blocks of pack ice. Five minutes later, to the
staccato sound of snapping timbers, its officers and crew took
to the whaleboats and rowed for the *Henry Taber*, where they
were taken aboard. The combined crews then watched in terrified silence as the doomed vessel's masts went over the side.
Moments later, its bow pointing skyward, the *Roman* slipped
below the gray waters stern first.

In years past the gamble had always paid off—and had
paid off handsomely. In midsummer the whaling fleet would
pass through the ribbon of shallow seawater known as the
Bering Strait and set a northerly course into the recesses of the
Arctic Ocean. Their quarry was the bowhead, a subspecies of
right whale that yielded an incredible 120 barrels of oil per animal. Once processed, the rich liquid became the source of
clean and odorless artificial light and served as a lubricant for
timepieces and the precision machinery of the industrial revo-

lution. The baleen, or whalebone, was fashioned into buggy whips and fishing rods, while ambergris, a waxy grayish substance formed in the beast's intestines, was used as a fixative in perfumes.

The hunt for *Balaena mysticus* lasted until late September or early October, just before the winter ice began to form. Because the weather pattern following the bowhead's discovery in 1848 pretty much held steady for the next twenty years, the temptation to prolong operations until the last possible moment became irresistible as the annual catch dwindled, an equally familiar occurrence in the world's other great whaling grounds. Their holds burgeoning with the spoils of the recent kill, thirty-two ships, including twenty-two from New Bedford, edged into the Bering Strait and the unbreakable grasp of a frozen sea.

The hunters were now the hunted as the ice ground toward the shore, trapping them in the ever-narrowing channel. The brig *Comet* was the next to crumple. Its crew abandoned ship, and the dashed hull and fittings were auctioned off on the spot for thirteen dollars. Two weeks later, with all hope lost of saving a cargo and ships valued at well over a $1 million, the captains ordered their 1,200 men into whaleboats for the perilous sixty-mile journey to Icy Cape. After three days and nights of rowing and tacking their way through heavy fog and blinding snow, they finally reached safety and were loaded aboard five barks. Remarkably, not one seaman had perished, but it was one of the blows from which the U.S. whaling industry would never recover.

Another had been struck by a former railroad conductor who had begun to exploit the riches of a vastly different sea. Judging by a photo taken in 1861, Edwin Laurentine Drake might have served as the double for the newly elected president from Illinois. Like Abraham Lincoln, Drake was angular and gaunt, sported a dark, scraggly beard, and wore a knee-length coat and stovepipe hat that made him appear considerably taller than he was. To further impress the good citizens of Titusville, Pennsylvania, which Drake described as a "dilapidated village" when he first arrived there in 1858, his employer, the Seneca Oil Company of New Haven, Connecticut, gave him the honorary title of colonel.

In the mid-1900s, Pennsylvania's Oil Creek was the best source of petroleum in the country. For centuries, the black fluid had been bubbling to the surface, where it was skimmed off by native Americans for use in medicines, vegetable dyes, and the weatherproofing of wigwams. The Seneca and other tribes passed the technique on to European settlers who had read Plutarch's account of how Alexander the Great, after conquering what is now Iraq, was impressed by the sight of a continuous flame issuing from the earth. Long before Alexander, the Egyptians had employed asphalt for embalming mummies and other purposes. More recently, the thirteenth-century explorer Marco Polo described oil and gas seepages near the Caspian Sea, while survivors of the de Soto expedition used oil from what is now Texas to repair their boats in 1543.

Drake's quest was spurred by the rising demand for oil for use in liniments, paraffin, and machine lubricants, all of which were dependent on the limited supply of whale and lard oil. Then, in 1854, the invention of the kerosene lamp led to the formation of the Pennsylvania Rock Oil Company, the firm from which Drake's employer leased land on Watson's Flat bordering Oil Creek. The price of petroleum shot up from seventy-five cents to two dollars a gallon. The fact that Drake owned a small amount of stock in the company deepened his resolve to succeed where a less tenacious speculator might have given up in disgust.

The thirty-nine-year-old venturer hired several locals to dig trenches and install a series of troughs and skimmers. The system was both cumbersome and inefficient, and it soon became apparent that the quantities of petroleum required to make a minor shareholder rich could never be produced by gleaning the surface of Oil Creek. Drake's thoughts turned to drilling, of which he knew nothing, except that salt drillers often had their wells polluted by oil. In 1824 they had accidentally hit oil in Kentucky and were run out of town because of the foul mess they made.

After months of frustration, during which he had no success locating a driller who did not "prefer Whiskey to any other liquid for a steady drink," Drake crossed paths with William "Uncle Billy" Smith, a blacksmith and experienced salt well driller who agreed to bore for $2.50 a day.

The sum also covered the labor of Smith's sixteen-year-old son, Samuel, the requisite drilling tools, including bits, and enough metal pipe to reach a depth of 400 feet. Drake added a six-horsepower steam engine and a "Long John" stationary tubular boiler with which to drive the drill into the ground, rotary drills being an invention of the twentieth century. His crew also built an enclosed wooden derrick and engine house, which reminded locals of a miner's shack with an unusually large chimney at one end.

Uncle Billy began pounding a chisel-like tool into the earth and scooping out the resulting debris. It was not long before gravel and groundwater began clogging the hole each time the driver was raised. Drake, who possessed the natural bent of an engineer, hit on a novel solution. With the aid of an oak battering ram, he had the crew drive ten-foot sections of cast-iron pipe down to bedrock thirty feet below. Drilling proceeded inside the metal casing, shielding the equipment from the unwanted fill.

On August 26, 1859, the drill bit suddenly dropped into a crevice. The workmen pulled the tools out of the hole, but because it was Saturday they opted to quit early and return to the rig on Monday. Though they had been drilling for weeks, the hole was only sixty-nine feet deep, too shallow, it was believed, for oil. In preparation for the coming week's labor, Smith and his son decided to ready the equipment late Sunday afternoon. When they peered into the pipe, they saw light reflecting off a dark liquid that seemed ready to spill onto the derrick floor. Samuel wheeled and bolted, shouting as he ran, "They've struck oil!"

The well was soon producing from ten to thirty-five barrels a day, almost doubling the world's production. Within a month other wells were being sunk along the banks of Oil Creek, giving rise to another nineteenth-century boomtown. Yet Titusville was unique, if only for the historical moment. Its wealth derived not from gold or silver, but from a liquid fossil fuel millions of years in creation.

Employing Drake's pioneering drilling methods, which the colonel, to his later regret, neglected to patent, speculators from across the country covered northwestern Pennsylvania with a forest of derricks. New towns bearing such prosaic names as Pithole City and Oil City blossomed in response to the frenzy. A year after the Drake well came in, oil was selling for twenty dollars a barrel. A year after that, the price had nose-dived to twenty cents, as hundreds of wells became thousands. Somehow the rate of discovery and pricing had to be brought under control.

In 1860 the world's first petroleum refinery was built about a mile from the now famous Drake well. Three years later the first pipeline, measuring two inches in diameter, began delivering about eighty barrels a day. The *Titusville Herald* and other local papers were publishing oil prices. Civic pride peaked with the founding of the Titusville Oil Exchange. Meanwhile, in nearby Cleveland, Ohio, a young dealer in hay, grain, and meats, sensing the commercial potential of Pennsylvania's expanding oil production, decided to build a refinery of his own. After its completion, John Davison Rockefeller would devote himself exclusively to the oil business.

The riches of which so many dreamed had suddenly become a reality for some, but most had failed, including hundreds that were scarred, maimed, or killed while working the frenzied fields. Three years after Drake's success, the colonel's employer was forced to sell its well because of depressed prices. Drake left Titusville for New York and promptly lost his money in the oil stock market. He eventually became an invalid and was reduced to begging for help in letters to his friends. Some looked on him as a flimflam man, a ne'er-do-well whose luck had inevitably run out, but members of the Pennsylvania legislature thought otherwise. In 1873 they voted Drake an annual pension of $1,500, which went to his wife, Laura, when he died bedridden in 1880. Still, the bogus colonel had lived long enough to see oil take its place next to coal as the second great energy source of the industrial revolution.

~

An observant Westerner traveling to such exotic lands as Peru, Egypt, or China in the 1880s would have noticed a reassuringly familiar product in the local markets and bazaars. It was the five-gallon kerosene tin stamped with the name of its manufacturer, Standard Oil. Not only did the corporation sell nearly all of the U.S. oil marked for export, it maintained a virtual monopoly on domestic production, shipment, and sales.

The genius and guiding hand behind this economic juggernaut were one and the same. Thirty years earlier, as he doggedly canvassed Cleveland's largest businesses in search of his first job, John D. Rockefeller could not guess what it would be. Yet, as he recounted in his memoirs, "I was after something big." Grave and meditative from an early age, the staunch Baptist was also patient and calculating. He finally landed a position with a firm of commission merchants trading with the railroads, steamship lines, and various manufacturers. From his meager salary of $3.50 a week, Rockefeller gave more than a biblical tenth to the Erie Street Baptist Church and helped a Negro freeman in Cincinnati purchase his slave wife, harbingers of the philanthropy to come.

Meanwhile, his attention had been drawn to the "oil regions" of Pennsylvania, where more and more petroleum was being pumped. This reckless surge in production was paralleled by the equally reckless construction of refineries by the score in the northeastern cities of Boston, New York, Philadelphia, Baltimore, and Pittsburgh. Indeed, if a man had access to a small amount of capital, it was no more difficult to open a refinery than to go into the livery business or to start up a hardware store. It was, however, a good deal more chancy as bust followed boom like a dog chasing its tail.

Despite the risks, Rockefeller's capitalist instincts convinced him that the "something big" he was searching for was now within his reach. Together with four associates, he built the Excelsior refinery on a three-acre tract near the Cuyahoga River in 1863. Within two years it was the largest refinery in the Cleveland area. Divesting himself of his other financial interests, Rockefeller bought out three of his fellow investors for $72,000, leaving only himself and Samuel Andrews, an expert in refining methods. "That," he later wrote, "was the day that determined my career."

Rockefeller soon brought his brother, William, into partnership and sent him to New York to develop the export and eastern trade. He built a second refinery, the Standard Works, and began to cast a covetous eye in the direction of his competitors. The high prices of 1865–66 triggered a jungle growth of small refineries—fifty in Cleveland alone—and Rockefeller saw his chance. When the economic cycle reversed itself a year later, resulting in a depression, he began purchasing the troubled refineries at a breakneck pace, backed by Cleveland's leading bankers. By the early 1870s, the Standard Oil Company, as it was now known, controlled nearly all the refineries in Cleveland, enabling Rockefeller to negotiate favorable shipping rates with the railroads and further crippling what remained of his competition.

With the railroads in his pocket, Rockefeller tightened the screws. Economists would later term his new strategy "vertical integration," the tentaclelike control of every resource needed for production and distribution. Standard Oil built its own cooperage plant for the making of barrels, purchased tracts of oak timber, manufactured its own sulfuric acid for refining, erected warehouses, created its own horse-drawn "tank wagon" service, acquired boats and tank cars, and procured the mineral rights to thousands of acres in Ohio and Indiana. Most important, the newly incorporated firm monopolized the system of delivery by sewing up the pipelines, the jugular veins of the refining industry.

It was a classic case of the new pseudoscientific doctrine of social Darwinism at work, which Rockefeller and his fellow industrialists embraced with the fervor of the anointed. As with life-forms inhabiting the wild, society must either adapt or become extinct, and no artificial impediment (meaning government regulation) must be allowed to interfere with the struggle for survival. The ideal society would be one presided over by a natural aristocracy of strivers—Rockefeller, John Pierpont Morgan, Jay Gould, Cornelius Vanderbilt, and, not least, Andrew Carnegie—industrialists and financiers whose habits of thrift and hard work are a model to others. Carnegie even added a pious religious gloss, standing traditional Christianity on its head by invoking what he called the "Gospel of Wealth:" "Those who would administer wisely" he wrote in 1889, "must, indeed, be wise, for one of the serious obstacles to the improvement of our race is indiscriminate charity." In celebrating the virtues of rugged individualism, the "captains of industry," as the historian Thomas Carlyle was the first to call them, also canonized ruthlessness, enabling them to exploit the uneducated, the downtrodden, and the environment

with an equable heart and a clear conscience, a self-styled variation of Manifest Destiny.

~

With Rockefeller firmly installed in the catbird's seat, the
drilling and refining of petroleum were at last brought into
balance. Refiners were able to sell every gallon of oil pumped
from the ground, except for the volatile lighter spirits produced in the first stage of distillation. These dangerous by-
products were normally discarded as waste, although some
found them particularly handy for cleaning stains off gloves.

A young Belgian metal enameler and inventor named Étienne Lenoir had another idea. He discovered that by pumping
air and "gasoline" into a small chamber, then setting off an
electrical spark, the mixture would literally explode. In 1860
the first version of the Lenoir gas engine was publicly un-
veiled. The basic design was similar to the steam engines of the
day but for the fact that the piston was driven by internal combustion rather than externally generated pressure. Some 500
of Lenoir's engines were built in Paris and another 100 in England.

Revolutionary though it was, the Lenoir engine exhibited
major drawbacks. Not only was the ignition system unreliable,
the explosive shock that occurred on firing caused many factory owners to fear that the engine might disintegrate at any
moment, showering their workers in shrapnel and flame.

Whether or not the visitors to the 1867 Paris exhibition
were mechanically inclined, few of them failed to take notice
of the nine-foot steel cylinder whose fluted exterior and large
pedestal base anticipated the streetlight of the 1920s. Atop the
cylinder were a number of gears, a series of cams and cogs,
and a large flywheel that together weighed over 1,600 pounds.
Concealed inside was a single piston, about six inches in diameter, driven by pressurized air and gas to the tune of thirty-six
strokes a minute. The blunderbuss generated about one-half
horsepower. (Also available was a larger two-horsepower
version weighing 4,450 pounds.) For being the best internal
combustion engine of its day, it was awarded the exhibition's
gold medal.

Its inventor, who had made a careful study of Lenoir's
engine, was Nikolaus August Otto, a dashing mechanical

engineer from Holzhausen, Germany. Having addressed and solved many of the engine's weaknesses, the bearded and mustachioed German moved to Cologne, where, with the aid of investor and technical adviser Eugen Langen, he opened N.A. Otto and Company.

It was Otto's good fortune to have hired Gottlieb Daimler, a gunsmith by training, as his technical director. Daimler in turn hired Wilhelm Maybach, who was destined to become one of the world's most renowned engine designers. The trio immediately set to work and in 1876 patented the Otto Silent Gas Engine, so named to set it apart from the convulsive rattletraps of the recent past.

Otto had always felt that the secret to developing a smooth-firing machine depended on the right combination of fuel and air. Utilizing Maybach's expertise in carburetion, he experimented with ever leaner mixtures to reduce the violence of the combustion. Once this was accomplished, he was faced with the problem of inconsistent ignition. Otto ingeniously hit on the idea of a "stratified charge." As the piston moves down during the intake stroke, it inhales only air during the first part of its movement, then an air-fuel mixture toward the end. The resulting explosion causes a mechanical linkage to operate a rotating cam, which pushes the piston back up the cylinder, expelling the burned gases. This all takes place in a series of four strokes: intake, compression, power (explosion), and exhaust, the classic Otto cycle. At 1,250 pounds and capable of generating two horsepower, the new engine weighed less than a third as much as its comparable predecessor, ran more smoothly, and burned less fuel.

In the meantime, Otto reorganized his company, renaming it Gasmotorenfabrik Deutz, and was soon turning out thousands of his four-stroke engines while amassing a fortune. Despite the fact that Daimler and Maybach profited as well, they decided to go their own way in 1882. Daimler, in particular, was convinced that Otto did not fully grasp the potential of the internal combustion engine.

The two set up shop in Bad Cannstatt, outside the flourishing industrial city of Stuttgart. Their objective was to develop a lightweight high-speed engine suitable for mounting on a vehicle. They first built an air-cooled, one-cylinder engine designed to run at 900 revolutions per minute, three times as fast as conventional engines of the time. They next built an even smaller motor and bolted it onto a wooden bicycle, which was tested successfully in November of 1885. One year later, the first Daimler four-wheeled road vehicle was rolled out of the shop—a carriage modified to be driven by a one-cylinder engine. This hybrid, which was quickly dubbed the "horseless carriage," is generally recognized as the first true automobile, although the German Karl Benz, who tested a three-wheeler powered by a two-cycle engine a year earlier, is perhaps deserving of equal billing.

Daimler's second model appeared in 1889 and incorporated major technological changes, giving it the look and performance of a true motorcar. The framework was made of lightweight tubing as opposed to solid wood, the engine was mounted in the rear, the wheels were driven by belts, and the driver had a choice of four speeds and steered with a tiller. Benz had already made his first sale to a Parisian in 1887, and Daimler was itching to get into production. The Daimler Motoren-Gesellschaft was founded in 1890, its business spurred when vehicles powered by Daimler engines finished first, second, and third in the wildly popular 1894 Paris-to-Rouen auto race. Daimler followed this victory by winning the first international motor race in 1895 and was soon building custom-designed race cars. His engines were also powering small boats, fire-engine pumps, and scaled-down locomotives, while Maybach developed the engine for the first Zeppelin airships. Within a decade or two, the internal combustion engine would become the largest single factor in making petroleum one of the most important energy sources in the world, enabling that dyed-in-the-wool Baptist John D. Rockefeller to become his country's first billionaire.

~

Not to be outdone by the British, the French sponsored a number of international expositions in the nineteenth century, one of which inspired the construction of the Eiffel Tower in the Champ-de-Mars, in 1889. At 984 feet, its iron framework rose more than twice as high as the tallest of the tall chimneys. Eleven years later, at the outset of the new century, the U.S. historian Henry Adams haunted the corridors of

Paris's Great Exposition until its doors closed in November of 1900. To the numbed and bewildered visitor, it seemed that the very universe itself was tottering.

Reflecting back in his famous essay "The Dynamo and the Virgin," Adams wanted to know "how much of it could have been grasped by the best-informed man in the world." As a Harvard graduate and professor, he was astonished by the complexities of the new Daimler motor and automobile, "threatening to become as terrible as the locomotive steam-engine itself," which was almost exactly Adams's own age.

No reactionary, Adams was well acquainted with the new technology and, even more unusual for a historian of his day, both wrote and lectured about it to his students. It was the velocity of change that had rocked him, the driving secularism of the new industrial order, especially as evidenced by his fellow Americans. These European transplants had never embraced the Virgin with the same fervor as their forebears, whose culture was deeply rooted in the Cross accepted by the Emperor Constantine prior to his death in 337.

It was to the Hall of Dynamos that Adams was repeatedly drawn, the place where mechanical energy was mysteriously changed into electrical energy. Thinking little of the "tons of poor coal hidden in a dirty engine-house carefully kept out of sight,...he began to feel the forty-foot dynamos as a moral force, much as the early Christians felt the Cross." What couldn't inventive human beings accomplish in the future? he thought to himself while standing before the huge wheel revolving at some "vertiginous speed" with barely a murmur. "Before the end, one began to pray to it; inherited instinct taught the natural expression of man before silent and infinite force." The continuity with the past had suffered an "abysmal fracture"; the once awesome planet Earth suddenly seemed far less impressive in its old-fashioned, deliberate revolution about the Sun.

Adams's dynamo has been compared to a new Prometheus, who, like his mythical predecessor, would prove himself a great benefactor to humankind. The industrial Titan would prove himself a creature of insatiable appetites as well. Not content with Earth's bounty of carbon and minerals, he had begun feasting on the sky.

Part Three
THE DWELLERS IN THE CRYSTAL PALACE

Whether we wish it or not we are involved in the world's problems, and all the winds of heaven blow through our land.

—Walter Lippmann, *A Preface to Politics*

NATIVE SON

~

But in science the credit goes to the man who convinces
the world, not to the man to whom the idea first occurs.

—Sir Francis Darwin, "First Galton Lecture"

On December 10, 1903, with King Oscar II and other members of the royal family looking on, four Nobel prize winners took center stage at the Royal Swedish Academy of Sciences in Stockholm. Conspicuous by their absence were the physicists Pierre and Marie Curie, who were represented by the French Ministry. The recipients stepped forward in turn to be addressed by Dr. H. R. Törnebladh, president of the academy, after which the king presented each with a diploma, a gold medal, and 141,000 kroner (about 40,000 U.S. dollars), from interest on the bequest of Alfred Bernhard Nobel, the Swedish chemist and inventor of dynamite.

The audience stirred when the name of Svante August Arrhenius was called, and for good reason. The forty-four-year-old professor of chemistry was the first native son to be so honored. Short, thickset, with a neatly trimmed blond beard and the fierce blue eyes of his Viking forebears, the laureate faced Dr. Törnebladh.

"Around 1880," the president intoned, when Arrhenius was in his early twenties and studying for his doctorate, he had traced the movement of electric current through various solvents, arriving at "a new explanation of the causes of chemical phenomena. The world of science already recognizes the importance and value of your theory, but its luster will continue to increase in the days to come. May your future work bear ever more abundant fruit and, when champions of the spirit and of learning advance along the trail that you have blazed, may your name be remembered in the proud words: *Ille fecit* [He did it]."

For a time it appeared that the academy's wish for Arrhenius would come true. The chemist was awarded both the Davey and Faraday medals; the universities of Heidelberg, Groningen, Oxford, Cambridge, Leipzig, and Birmingham all granted him honorary degrees; his many books sold in the thousands, capturing the public's imagination with titles that sounded more like the works of a romantic poet or a master of science fiction: *The Destinies of the Stars, Worlds in the Making, Life in the Universe.*

Yet within a few years of his death in 1927, at age sixty-eight, Svante Arrhenius was all but forgotten outside Scandinavia. Only in the 1990s would his memory be rekindled because of an obscure paper he had published a century earlier. Within its pages Arrhenius set forth the startling idea that the massive consumption of fossil fuels is capable of raising the temperature of the atmosphere. For the beginnings of that fascinating story, we must briefly return to Arrhenius's graduate school days and the young chemist's bitter struggle to secure himself a permanent place in the academic community.

~

The renowned German chemist and Nobel laureate Friedrich Wilhelm Ostwald well remembered the day in 1884 when, simultaneously, he became the father of a baby girl, contracted a nagging toothache, and received a doctoral dissertation in the mail from young Arrhenius, a newly minted Ph.D., of whom Ostwald had never heard. Years later, after claiming his own Nobel prize, Arrhenius quipped, "The worst was the dissertation, for the others developed quite normally."

Arrhenius's reason for contacting one of the luminaries of modern physical chemistry was straightforward enough: When he had presented his dissertation of the University of Uppsala weeks earlier, it was awarded only a fourth class; that is, "approved without praise." He fared only marginally better in his oral defense, which garnered him a third—"approved with praise." This poor showing disqualified the young Swede from receiving a university lectureship, leaving him deeply humiliated and fighting mad.

As Arrhenius later told the story, it was on the evening of May 17, 1883, that he experienced one of those eureka moments in science. "I could not sleep that night until I had worked through the whole problem." His insight yielded the theory of electrolytic dissociation. It explains that those substances that, when dispersed in any solvent, are good conductors of electricity are also those substances that, in solution, decompose into atoms carrying powerful charges known as ions. The next morning, the disheveled but ebullient graduate student made directly for the office of his chemistry professor Per Teodor Cleve. "I have a new theory of electrical conductivity as a cause of chemical reactions," he blurted out. But before he could explain further, he was interrupted.

"This is very interesting," Cleve replied. Then he abruptly dismissed an astonished Arrhenius with a resounding "Goodbye!"

Only when Arrhenius was tapped for the Nobel prize twenty years later did Cleve see fit to explain himself. Scientific theories are constantly being formed, he noted, but almost all of them turn out to be wrong. By applying statistics to the formulation of ideas, Cleve had concluded that Arrhenius's theory would abruptly vanish as well, so why even bother to discuss it.

Like others to come, Cleve greatly underestimated this descendant of tough-minded farming stock. One Lasse Olofsson, a remote ancestor, had settled in the village of Arena in 1620, from which the family derived its Latinized surname, Arenius, the spelling of which was changed in the early nineteenth century to its present form. Arrhenius's father, Svante Gustave, attended the University of Uppsala, later becoming surveyor of the city and a collector for the university.

The second of two sons and his father's namesake, Svante August was born on February 19, 1859. The boy's intellectual gifts manifested themselves early on. Like the philosopher and economist John Stuart Mill, he taught himself to read at the age of three, while his unusual mathematical skills were stim-

Svante August Arrhenius
(Used by permission of The Granger Collection, New York)

ulated by observing his father add up columns of figures in the university account books. When he was old enough, the stoutish, square-jawed youth was enrolled at the Cathedral School of Uppsala, where he was taught physics by its rector, M. M. Floderus, who was also the author of his text. Having compiled an excellent academic record, he entered the University of Uppsala at the age of seventeen and soon passed the candidate's examination, admitting him to study for the doctorate.

Wilhelm Ostwald was not the only recipient of Arrhenius's dissertation. The frustrated scholar dispatched copies to other prominent scientists in Bonn, Amsterdam, and Tübingen. Meanwhile, Otto Pettersson, a professor of chemistry at the Högskola, Stockholm's technical university, wrote a review of Arrhenius's monograph in which he came to a rather different conclusion than that of the Uppsala professorate. "The faculty have awarded the mark *non sine laude* [not without praise] to this thesis. This is a very cautious but very unfortunate choice. It is possible to make serious mistakes from pure cautiousness. There are chapters in Arrhenius's thesis which alone are worth more or less all the faculty can offer in the way of marks."

Ostwald was of a similar mind and decided to take the measure of the upstart scholar himself by paying a visit to Uppsala. The two chemists hit it off immediately; both men loved their beer almost as much as they loved science, and each polished off a minimum of three large glasses at a sitting. Although Ostwald's presence turned some heads in the local academic community, no offer of a position was forthcoming. A peeved Ostwald insisted that Arrhenius join him in Leipzig, where he promised to set aside part of his private laboratory for the young man's use.

Thus began the *wanderjahre* (1886–90), during which Arrhenius worked with Ostwald in Leipzig, Friedrich Kohlrausch in Würzburg, Ludwig Boltzmann in Graz, and, finally, Jacobus H. van't Hoff, the first Nobel prize winner in chemistry, in Amsterdam. Like his German and Austrian hosts, the notoriously stiff Dutch were quickly won over by their visitor's sharp wit and irresistible charm. He was "Dear Svante" to van't Hoff and the employees who worked in the laboratory.

During his nearly four years of "wandering," Arrhenius expanded on his theory of electrolytic dissociation, which was beginning to gain grudging acceptance among academics. At the same time, the noted chemists who had made him their collective protégé never for a moment left the authorities back in Sweden in any doubt about their feelings.

Finally, in 1891, Arrhenius was offered a post in his native land, that of lecturer in physics at the Högskola. He was elevated to a professorship at the same institution in 1895. A year later, an unusual article appeared under his name in one of the still prolixly titled scientific journals: *The London, Edinburgh, and Dublin Philosophical Magazine and Journal of Science*. The paper, which had been presented to the Royal Swedish Academy of Sciences the previous December, was titled "On the Influence of Carbonic Acid in the Air upon the Temperature of the Ground."

Here Arrhenius sets out to answer the following question: "Is the mean temperature of the ground in any way influenced by the presence of heat-absorbing gases in the atmosphere?" Exactly how and when he was first attracted to this idea he does not say. Yet he readily admits that it is not original with him. "Fourier," he writes, "maintained that the atmosphere acts like the glass of a hothouse, because it lets through the light rays of the sun but retains the dark rays from the ground." He had also studied the work of the Irish polymath John Tyndall, an avid mountain climber and student of glacial action who began his career as a surveyor and engineer in England during the railway mania of the 1840s. In a paper published in 1861, Tyndall demonstrated that various gases in Earth's atmosphere—aqueous (water) vapor, carbon dioxide (CO_2), and ozone—possess high absorbent powers, which he succeeded in measuring. While he reasoned that a significant drop in the amount of CO_2 in the air would likely result in another ice age, he never considered the other side of the coin—that human intervention might trigger global warming.

The work of a U.S. astronomer also caught Arrhenius's eye. In July of 1881 Samuel P. Langley, the noted solar physicist and director of the Allegheny Observatory, led an expedition up the slopes of Mount Whitney in southern California to conduct research on solar heat and its absorption by Earth's atmosphere. The project was partially financed by the Signal Corps, a branch of the U.S. War Department, under whose auspices Langley published the results in 1884.

Langley observed that during the ascension, as the air grew thinner, the temperature fell. "We might infer . . . that if the air grew rarer still, the temperature would fall still more, and that *when the air was altogether absent, the temperature of the Earth under direct sunshine would be excessively low.*" Just how low was a matter of calculation. Employing a formula reprinted in his page notes, Langley again resorted to italics: *"the temperature . . . would probably fall to –200ºC if [the] atmosphere did not possess the quality of selective absorption."* After a later study of the Moon, whose atmosphere has long since dissipated due to a lack of gravity, Langley realized that he had, in Arrhenius's words, based his calculations "on too wide a use of Newton's law of cooling." Still, the astronomer's fundamental premise was correct: Remove Earth's heat-absorbing gases and the planet would be locked in icy thrall.

Although this area of inquiry was far removed from the chemical research that would gain him the Nobel prize, Arrhenius was keenly aware of the changes being wrought in the environment by rampant industrialization. His native Uppsala, which became the pagan capital of Sweden in the sixth century, was now a major railroad junction and manufacturer of metal goods and textiles. He had seen the giant smokestacks of a northern Europe in the throes of a technological revolution and had often entered into lively discussions on the subject with friend and colleague Professor Arvid Högbom, both at the Högskola and at meetings of the Physical Society of Stockholm.

Sporting a handlebar mustache and thick wavy hair that he parted down the middle, the urbane Högbom, a geologist by training, was concerned with the natural processes that may have caused variations in the atmospheric level of CO_2 during different geologic cycles—most particularly the disappearance and return of the great ice ages. In the spring of 1893, he delivered a lecture to the Physical Society that Arrhenius largely incorporated into his own article in the *Philosophical Magazine*. In the lecture Högbom observed that the chief source of Earth's CO_2 is the release of the gas during the natural breakdown of limestone. The organic world of plants provides additional quantities, and a small but seemingly inconsequential amount is derived from the annual burning of 500 tons of coal by industry. As to how much CO_2 is retained by the atmosphere, how much is absorbed by the seas, and how it might effect climatic change on a global scale, Högbom could only theorize.

Additional impetus had been provided by a natural cataclysm that took place in 1883, when Arrhenius was on the verge of completing his doctorate. In southwestern Indonesia, in the Sunda Strait between Java and Sumatra, lay the tiny island of Palau. Its dominant geologic feature was the long-dormant volcanic mountain Krakatau, rising 2,600 feet above the sea. The only recorded eruption had occurred more than two centuries earlier, in 1680, and had caused relatively minor damage.

On May 20, 1883, one of Krakatau's multiple cones suddenly became active, spewing ash-laden clouds 6 miles into the atmosphere. The explosions were heard in Batavia 100 miles distant, but within a week the activity had died down. Then, at 1 P.M. on August 26, the first of a series of increasingly violent explosions occurred, sending a black cloud of ash 17 miles into the sky over the Java Sea. The climax was reached at 10 A.M. the next day, heralded by a series of tremendous explosions heard 2,200 miles away in Australia, and thought to be among the loudest sounds in history. Ash was propelled 50 miles into the atmosphere, while pressure waves were recorded around the globe.

By the time it was over the following day, Krakatau had cast 5 cubic miles of rock fragments and vast quantities of ash over an estimated 300,000 square miles. Near the volcano itself, masses of floating pumice swarmed so thickly that ships attempting to enter the Sunda Strait were forced to turn back. Night had suddenly fallen on the morning of the eruption, and the Sun would not be seen again for two and a half days.

Borne by what are now termed Krakatau winds, fine volcanic dust particles circled the planet time and again, diffusing the Sun's light and triggering alarm bells that sent horse-drawn fire wagons racing toward the "flaming" horizon. These same vivid sunrises and sunsets inspired the aesthetically sensitive, who commemorated them in paintings and poetry.

Only six square miles of a decimated Palu—one-third its former area—remained next to a basin covered by 900 feet of seawater. On the nearby islands of Verlaten and Lang, ash and pumice had accumulated to a depth of 200 feet, annihilating all life. Tsunamis triggered by the volcano's collapse were recorded as far away as South America and Hawaii. The greatest of the tidal waves, 120 feet high and moving at an incredible 100 miles per hour, claimed an estimated 36,000 lives as it engulfed the coastal towns of nearby Java and Sumatra.

After the 1783 eruption of a volcano named Laki in Iceland, Benjamin Franklin had theorized that aerosol clouds—gaseous suspensions of fine particles—in the upper atmosphere scatter sunlight, reducing the level of solar radiation reaching Earth's surface. If sufficiently widespread, these gases would tend to lower the planet's temperature, although there was no way to prove it at the time. Krakatau's eruption led Högbom to conjecture that a series of similar events in ages past had combined to trigger the periodic advance of the glacial ice. Yet as Arrhenius himself would later write in *Worlds in the Making*, there was an analogy to be drawn between the sudden release of Krakatau's fury and the rapid expansion of industrial pollution: "The more volatile constituents, such as carbonic acid, sulphuretted hydrogen, and hydrochloric acid, may spread over large areas and destroy all living things by their heat and poison."

~

It was during a long period of personal discontent that Arrhenius set about formulating the first theoretical model with which to calculate the influence of carbonic acid (CO_2) on the temperature of Earth. Separated from his beautiful and intellectually gifted wife, Sofia Rudbeck, who had been his best student as well as his laboratory assistant at the Högskola, he reluctantly consented to a divorce. Custody of their only child, a bright boy named Olof, was granted to the mother, and Arrhenius would not be allowed to visit his son until Olof had reached the age of five. In an attempt to keep his loneliness and depression at bay, the chemist gathered stacks of scientific journals, reams of writing paper, and numerous pens. Then, on Christmas Eve of 1894, he rolled up his sleeves and began what he later described as the most "tedious calculations" of his life.

Rising long before a wan Nordic Sun made its brief appearance and retiring well after midnight, Arrhenius put in fourteen-hour days. "I have not worked this hard," he wrote Ostwald, the one person with whom he shared his private tribulations, "since I was cramming for my B.A." At first numbering in the tens, then the hundreds, and ultimately the thousands, the equations, which today can be done by a computer in split seconds, filled countless pages with mathematical symbols reminiscent of ancient runes—perpendicular, oblique, with the occasional squiggle for good measure.

Winter turned to spring, spring to summer, and summer to autumn, and still Arrhenius labored on. The seasonal round had come full circle when he wrote his good friend and fellow scientist Gustav Tamman, "It is unbelievable that so trifling a matter has cost me a full year."

When he finished at last, his purely theoretical model was far in advance of anything contemplated by Fourier or Tyndall. Where Fourier had envisioned the atmosphere as a bell jar lined with black cork "designed to receive and conserve heat," Arrhenius was the first to employ the term *hothouse*, which would be renamed the *greenhouse effect* decades later.

The chemist saw it this way. Two gases—water vapor and CO_2—are responsible for the warming of Earth's atmosphere. Because they are transparent to light emitted by the Sun, its rays pass through and strike the planet. The light is largely absorbed and reemitted at Earth's surface as infrared radiation, or heat, invisible to the naked eye. Due to its altered wavelength, the infrared radiation cannot pass back through the water vapor and CO_2 as before. Some of it is absorbed, and part of what is absorbed is radiated back toward Earth's surface, thus trapping heat and warming the planet. If Arrhenius was right, any significant and prolonged alteration in the concentration of these gases could trigger major climatic changes. Reduce the amount of CO_2 in the atmosphere, and the raging winter—"wind time, wolf time" of the Norse sagas—might well return; raise the amount of the gas, and the northerly latitudes would likely burgeon and blossom as never before.

Arrhenius's herculean calculations enabled him to construct a series of mathematical tables. Number seven, described by his recent biographer as his "pièce de résistance," is titled "Variation of Temperature Caused by a Given Variation of Carbonic Acid." The author provides seasonal differences in Earth's temperature from 70° north latitude to 60° south latitude based on CO_2 levels ranging from 0.67 to 3 times the estimated amount in the mid-1890s. He concludes that a drop in the level of CO_2 to some 0.62 to 0.55 of its present value would result in a temperature lower by four to five degrees Celsius (7.2° to 9°F), a change sufficient to trigger another ice age. Conversely, a doubling of CO_2 would lead to an average temperature increase of five to six degrees Celsius (9° to 11°F). For the time being, he was betting on fire over ice due to the rising consumption of fossil fuels and the release of heat-absorbing CO_2 into the atmosphere.

Why, exactly, did Arrhenius immerse himself in these "tedious calculations" to resolve what he characterized as a "trifling matter" in his letter to Tamman?

While one cannot discount the possibility that a degree of false modesty was involved, the answer would seem to lie elsewhere. Scientists are professional puzzle solvers—first, last, and always—yet the specific nature of the puzzle is not necessarily their primary concern. Rather, they are intrigued by how best to formulate the most accurate solution possible. Even as Arrhenius was complaining about the inordinate amount of time it was taking to complete his calculations, he also informed Tamman that he would have abandoned the pursuit long since had it not been for "an extraordinary interest" in the problem.

Add to this a seeming lack of urgency. Arrhenius could only extrapolate from the limited data at hand, and it pointed him in the direction of Darwin rather than Cassandra. In the version of his paper presented to the Swedish academy, he estimated that it would take another three millennia of burning fossil fuels for the amount of atmospheric CO_2 to double. This did not seem an especially alarming prospect to a thick-blooded Scandinavian, who passed much of the year in winter twilight. Indeed, the more he pondered the idea of atmospheric warming, the more it appealed to him. The projected increase in temperature, he informed those who attended his lecture, will "allow all our descendants, even if they only be those of a distant future, to live under a warmer sky and in a less harsh environment than we were granted." Several years later, in 1908, he echoed this prediction in *Worlds in the Making*: "By the influence of the increasing percentage of carbonic acid in the atmosphere, we may hope to enjoy ages with more equable and better climates, especially as regards the colder regions of the Earth, ages when the Earth will bring forth much more abundant crops than at present for the benefit of rapidly propagating mankind."

10

"NEVER A MAN"

~

One generation passeth away, and another
generation cometh: but the earth abideth forever.
The sun also ariseth.

—Ecclesiastes

In a remote valley of northeastern Arizona, where its borders meet those of Utah, Colorado, and New Mexico to form the Four Corners, archaeologists brought in a backhoe to cut trenches through an ancient village site called Kin Klethla. As it was gouging a path across a series of burned-out rooms, the steel-toothed bucket unearthed a single human skull, minus the mandible. The forehead of the skull, thought to be that of a female, has a gaping circular hole, a nearly perfect match to the empty sockets whose bright dark eyes had once gazed upon the surrounding valley before they were extinguished by a single blow. In addition to the symmetrical fracture, two large cut marks traverse the skull, as if the slayer had taken no chances. Archaeologists now feel certain that the death and burning occurred in the second half of the thirteenth century, when Kin Klethla and hundreds of other villages like it were suddenly abandoned forever.

Although this vanished culture had once encompassed an area the size of New England, the scope of its existence remained unknown until the late nineteenth century. Near Christmas in

1888, brothers Richard and Al Wetherill, together with their brother-in-law, Charles Mason, were searching for stray cattle during a snowstorm in southwestern Colorado. The cowboys rode out of a canyon and onto a mesa in hopes of gaining a better view. When they paused to look across Cliff Canyon toward Mesa Verde, or "green table" as it had been named by early Spanish explorers, they could just make out the shimmering image of great buildings seemingly suspended in the void. An hour or so later found the trio climbing down into a large natural recess halfway to the bottom of a steep cliff. There they entered a magnificent pueblo composed of multistoried masonry buildings. Inside the deserted rooms were fine pottery, stone tools for grinding corn, and the ashes of fires that had not glowed for 600 years. The astonished visitors named the lost city Cliff Palace, which together with another 5,000 sites in Mesa Verde National Park has become the destination of tens of thousands of tourists each year.

~

Called the Anasazi, or "Ancient Ones," by their modern descendants, they were the distant offspring of nomadic Asian hunters. For thousands of years they trekked across the land bridge that once connected Siberia and Alaska, driven by arctic winds streaming southward over the great ice. The first of them reached the arid, pine-clad Four Corners as early as 10,000 B.C., but it was not until the first millennia before Christ that they adopted agriculture and began making the distinctive pottery whose shards lie scattered by the countless thousands across the desert floor and high mesas.

The Anasazi bore the genetic imprint of their Asian ancestors. Dark-eyed with black hair, high cheekbones, and brown skin, they were squarely built, the men averaging five feet, four inches in height, the women two inches less—about the same as Europeans of the time. Their clothing consisted of robes of hide, fur, feathers, and woven cotton, sewn together with bone needles and fibers from the stout-stemmed yucca plant; finely spun dog hair occasionally served as a sash. Ornamentation took the form of copper bells from Mexico and seashells from the Pacific Coast. In return the Anasazi offered highly prized turquoise, which they mined at a site over a hundred miles away.

When times were good, they dined on game, corn mush, domesticated turkey, squash, pine nuts, wild vegetables, and dried buffalo supplied by the Plains Indians farther east. Flat cornmeal cakes were baked on stone griddles, while sacred corn flour was offered to the gods. In lean years they relied on the surplus set aside to see them through—but mostly they prospered, for the spirits were kind. Even the crippled and the elderly abided, and were taken care of long after they could no longer work. When the end came at last, the dead were buried with tools and adornments that would serve them in the after-life.

Anasazi culture reached its zenith in the eleventh century A.D., the height of the European Middle Ages. The region's population doubled as agricultural production expanded to both higher and lower elevations, spurred by abundant rain-fall and moderate temperatures. The period was accompanied by an explosion in archaeological evidence known to those fa-miliar with the scientific literature as the "Chaco Phenome-non," so named for the people and the valley they occupied in what is now northwestern New Mexico. There the Anasazi built the largest of their rectangular Great Houses, multilevel "apartment" buildings in which scores of families each had their own suite of rooms, some measuring twelve feet long and eight feet high. The walls were finished with white plaster and decorated with frescoes, while the windows and doors were kept small to retain heat in winter. Warmth was provided by interior wood fires that blackened the ceilings, for there were no chimneys at Chaco—nor anywhere else in the Anasazi world.

The Great Houses also contained large circular structures known as kivas. These were entered by a single ladder that led down to the floor through a hole in the beehive-shaped roof. Warm in winter and cool in summer, kivas were used for reli-gious rites, weaving, toolmaking, storytelling, and family gath-erings. The participants sat on a stone bench that circled the wall at a respectful distance from the center of the room, where a sacred hole in the floor permitted spirits to exit the underworld.

By 1050 Chaco Canyon, fifteen miles long and a mile wide, contained twelve elaborate towns that had blossomed

Cliff Palace, an ancient Anasazi site
(Used by permission of Index Stock Photography, Inc.)

into the religious, political, and economic center of Anasazi life. Like spokes in a wheel, paved roads radiated outward, connecting the canyon to many distant communities in a pattern replicated on a lesser scale throughout the Four Corners. The people of the Chaco, numbering perhaps 5,000, had fashioned the most complex prehistoric American Indian culture north of Mexico.

For all its brilliance, the world of the Ancient Ones was existing on borrowed time, a civilization in precarious equipoise between survival and oblivion. Soon after A.D. 1100, a winter drought set in. This in itself was nothing new, for experts in the science of dendrochronology—the study of the growth rings in trees to determine the date of past events—have documented many such cycles in the prehistoric Southwest. This time, however, the normally dependable summer monsoon arrived later and departed earlier, compounding the problem. The Anasazi compensated by employing a number of water control strategies, including check dams, ditches, and reservoirs. Eventually the drought passed with few apparent consequences, and life returned to normal.

That it did so is explained by the fact that the Anasazi inhabited an era known as the Little Climatic Optimum or Medieval Warm Period, which stretched from A.D. 900 to 1300. Temperatures during this time were slightly warmer and rainfall was plentiful enough to support the steady expansion of agriculture. In Europe vineyards were established five degrees farther north and grew 200 meters higher above sea level than in the 1960s. Strange as it seems, the region around England's river Tyne, which forms in Northumberland, could have been known for its local vintage.

Beginning around A.D. 1213, signs of a major climatic change started to develop. Colder and drier air moved into the Southwest, making the farming of the sacred corn ever riskier at higher elevations. This subtle cooling of the atmosphere was accompanied by summer and winter drought that hit the farming belt from the bottom. It was as if a pair of giant hands were pressing together from above and below, shrinking the arable land by the season. After a decade of this, conditions improved once again, doubtless leading to a false sense of security. But as we now know, the Little Climatic Optimum had begun to play out, and was soon to take with it the halcyon days of Anasazi memory.

The "land of little rain," as the writer Mary Austin described the desert Southwest, became a land of lesser rain after 1250. For twenty-seven years (1271–97)—the span of a generation—the dwellers in the Four Corners suffered as they never had in the preceding five centuries. The shrinking tree rings of the piñon, bristlecone, and ponderosa pine, as well as the desiccated spruce, foretell the tragic story of what archaeologists refer to as the Great Drought.

The acquisition and preservation of food, always of utmost concern, now became an obsession. The cooler temperatures and unpredictable rainfall not only affected annual yields but resulted in the depletion of other critical resources such as firewood, wild plants, and game animals, further hastening the downward spiral. As in the tree rings, the story is inscribed on bone. Skeletons and teeth found in the ruins provide evidence that the Anasazi suffered malnutrition, shorter life spans, and a rise in infant mortality. Driven by the cries of their starving children, the Anasazi resorted to the club and the arrow.

The plundering of Kin Klethla that resulted in the savage death of the young woman by braining was acted out many times in the parched valleys of the Four Corners. At the same time, the beleaguered residents were expending ever more energy by constructing dams and canals to trap and divert water to their terraced fields. Finding it impossible to work the land and protect their families and meager stores at the same time, they opted for the security of the cliffs, where they constructed new villages in natural caves accessible only by woven ladders or chiseled handholds. Here, on slanting floors, they perched like wary birds—the elderly marooned forever, the young in constant danger of toppling onto the jagged rocks below. Water, their most precious resource, was carried up in geometrically lined pots of Kayenta black on white, the shards from which serve as reverse clocks, ticking backward to the waning life of the mesas.

Even the security of the natural recesses was not enough. Incredulous archaeologists have recently discovered village sites atop the very mesas themselves. Such is Six Foot, a cliff island named for its still-upright wall. Reachable only by a crack wide enough for a single individual to traverse at a time, the village of seventy-five to one hundred souls could have easily been defended from would-be raiders by an old woman armed with a pile of stones. What is more, Six Foot was part of a network of similar villages that could watch out for and alert each other of potential danger, whether by smoke signals or some type of reflective device. Still, this new way of living was itself as fragile as a house of cards. If one strategic site succumbed, the others were liable to become part of a domino effect, each collapsing in turn.

Suddenly, they were all gone—every man, woman, and child. The prolonged drought had baffled even the wisest of the Ancient Ones, whose gods had failed them, or was it the other way around? So far as is known, not a single Anasazi remained in the land of his ancestors by the year 1300, which climatologists mark as the official beginning of the Little Ice Age. All it had taken was a sustained drop in the yearly temperature of as little as one degree Celsius accompanied by an annual decline in rainfall of a little over an inch, a mere shrug of eternity.

~

In 1361, a few decades after the last of the Anasazi departed the Four Corners, Ivar Bárdarson, a seafaring Norwegian priest, and his companions sailed up Greenland's west coast. They were hoping to make contact with their fellow Christians, from whom nothing had been heard for several years. But on reaching what the Norse called the Western Settlement, they found only some untended cattle and sheep. As Bárdarson chillingly recounted, "There was never a man."

More than three and a half centuries earlier, the Vikings had first reached Greenland in their long ships, colorfully described in their poetry as "oar-steeds," "surf-dragons," and "fjord-elks." They were led by the half-mad exile Erik the Red, so named for the tint of his wild mane and the blood on his hands three times over.

Born Erik Thorvaldson in the tenth century, the future explorer left his native Norway for Iceland together with his father, a tough customer named Thorvald Asvaldsson, after the two had been banished for manslaughter. Once in his new home, Erik married into a good family and set himself up as a farmer in the valley of Haukadalur. It did not take long, however, for his violent streak to resurface, and he was twice more found guilty of multiple killings. With Norway and now Iceland barred to him for the next three years, he decided to sail west in hopes of locating and exploring a new land sighted some fifty years earlier by one Gunnbjörn Ulfsson, a Norwegian sailor driven off course during a great storm.

In the year 982, heading due west along what is now the sixty-fifth parallel, Erik and his shipload of followers, livestock, and household goods safely negotiated the 175 miles of treacherous seas, only to strike a forbidding coast so choked with drift ice that a landing was impossible. Having little choice, the party altered course and followed the coastline southward until they rounded the tip of the vast island continent. Though rough and rugged, the west coast of Greenland was more inviting than its bitterly inhospitable eastern shore. In the folds of the icy mountains were deep fjords, well protected from the surging, fish-laden sea and warmed by the Gulf Stream. Game in the form of caribou, seal, walrus, and polar bear abounded, the furs and ivory and hides from which would command high prices on the European market. Most important, pockets of green fertility, seemingly well suited to the dairy farming left behind in Iceland, separated the coast from the endless ice cap, stretching ever northward to the spinning axis of the world.

No stranger to homesteading, Erik plied the western coast, searching out the most promising areas for colonization. As First Settler, he was entitled to claim the best land for himself, which he did without embarrassment. He and his stoic companions then endured three long winters, the arctic Sun slowly ticking off the remaining days of his sentence. At last, his banishment over, Erik returned to a starving Iceland with a message of hope and opportunity. As befitted his character, he spoke of a wondrous land over which the Sun sets in the evening, a place he had cleverly named Greenland, because, as the sagas relate, "people would be the more eager to go there if it had a good name."

In the early summer of 986, Erik set out at the head of the "fleet of the otter's world," a flotilla of twenty-five ships packed with 300 desperate but hopeful emigrants and all their worldly goods. If the *Greenlander's Saga* is to be believed, only fourteen vessels reached their destination on Greenland's southwest coast; the rest either foundered in "sea-fences"— towering waves—or came to grief in rending encounters with the drifting ice.

From this tenuous beginning two colonies were established. Erik himself chose a site near the southern tip of Greenland known as the Eastern Settlement, the nucleus of which was his farm called Steep Slope. Its grassy hillsides tumbled down to an intense green plain on the shimmering fjord Tunugdliarfik, the pull of whose waters would one day tempt his restless son, Leif the Lucky, to point his long ship westward toward a previously unexplored coast called Vinland. By the middle of the twelfth century, the Eastern Settlement boasted a population of between 3,000 and 5,000 souls. Having forsaken their Christless chivalry for the Cross, they traded a live polar bear for a Norwegian bishop in 1125. As a further inducement they built a cathedral of stone fitted with imported stained-glass windows and bronze bells. The bishopric was supported by a church-owned farm that grew to include two-thirds of the best grazing land on the island.

Some 170 miles to the north lay the smaller Western Settlement, whose population never exceeded 1,500. Their nearest neighbors were the heathen Skrälings—in medieval Norse the "ugly ones" or "savages." A nomadic people, the Inuit or Eskimo, as we know them today, relentlessly tracked their food across the ice, living anywhere their quest for seal, walrus, caribou, and polar bear might take them.

By contrast, the Norse were committed subsistence farmers. They had brought with them in their open ships shaggy cattle, sheep, and long-haired goats, from which came the meat and milk that were central to their diet. Each autumn they harvested and stored the wild grasses to feed their livestock, which were wintered over in byres of heaped earth and stone, lest they perish in the wind and cold. The Norse hunted caribou as well, accompanied by great hounds possessed of thick grayish coats with tails that curled up over their backs. Seals were taken during the pupping season when the vulnerable females and their offspring were confined to land, but there is little evidence to suggest that fishing was important to the economy, for few bones and no gear have been unearthed. Nor, apparently, did domesticated plants contribute to their diet, although wild fruits such as the crowberry were gathered in the summer.

The wood used for cooking and heating was carried to shore by the ocean currents, while building timbers were imported from Europe as part of a vigorous trade that lasted the better part of four centuries. In return for metal, woven cloth, and tools, the Norse offered walrus ivory, polar bear skins, and other so-called prestige goods, some of them likely acquired in trade with the resourceful Inuit.

Even more than the contemporaneous Anasazi, the Norse teetered on the verge of catastrophe from beginning to end. No matter how plentiful the harvest, there always came a time of maximum vulnerability in late winter when it seemed as if the dwindling supply of hay would never see the livestock through. Then, beginning about 1300—the time of the Anasazi's demise in the Four Corners—the ice returned in full fury.

The onset of the Little Ice Age saw average temperatures in the North Atlantic region drop by about 1.5 to 2 degrees Celsius. The growing season gradually turned both cooler and shorter, placing pressure on the hitherto thriving Norse. Vital trade with Iceland and Europe, which had already begun to dwindle, now ceased completely for long periods as the ice-girded sea blocked traditional sailing routes. Somewhere between 1350 and 1355, the Northern Hemisphere experienced its most frigid winter in 800 years. And this was almost exactly in the middle of the fourteenth century's longest cold spell, which lasted from 1343 to 1362. The seals no longer came to pup, and the walrus and polar bear were inaccessible by boat. Inevitably, this twenty-year clustering of poor summers led to what the later colonists of Jamestown came to know as the "starving time." Little wonder that it dovetailed perfectly with Bárdarson's account of the Western Settlement's utter decline.

As was true of the Anasazi, the tribulations of the Norse were likewise inscribed on tooth and bone—both human and animal. Viking teeth recovered from 500-year-old corpses not only bespeak malnutrition and disease, but also provide hard scientific evidence of climatic change. Tooth enamel records the ratio of oxygen-18 to oxygen-16 atoms in water consumed by children. Oxygen-16 is the lighter of the two isotopes and evaporates more readily from the oceans to form clouds, which in turn produce snow. The fewer the oxygen-18 atoms present, the colder the temperature of the atmosphere. By comparing children's teeth from the year 1100 with those of 1450, scientists at the University of Michigan found that the mean annual temperatures had dropped by about 1.5 degrees Celsius, exactly in line with the results produced by other dating techniques.

Further testimony has been provided by skeletal remains from Norse graveyards. As conditions became more brutal and food grew ever scarcer, the average height of male Greenlanders declined from about five feet, eight inches in Erik the

Red's time to about five feet, four inches by the beginning of the 1400s.

Some of the most poignant evidence that all was lost came to light when excavators working at the Western Settlement unearthed the dining areas of four homes. In each they discovered the remains of the large elkhounds used in the hunting of caribou. Several bones of the great canines bear cut-marks, strongly suggesting that the Norse sacrificed one of their most favored possessions in return for a few more days of life, which was about all the wasted flesh of the beasts—starving to death themselves—could provide. That the fodder had run out and the cattle had already been consumed is attested to by the presence of inedible hooves. In their desperate search for calories, the Norse had literally eaten themselves out of their subsistence.

It was a tragedy some would argue need never have happened. For even as the Norse were dying off or taking flight, the wandering Inuit thrived. Rather than follow the example of their shamanistic neighbors, the colonists fought to preserve their way of life to the bitter end.

To begin with, there was the clothing. The Inuit protected themselves from the penetrating cold by fashioning sealskin parkas and trousers lined with the skins of migratory birds. The Norse kept to woven garments of the type popular in medieval Europe. In the local churchyards the women were buried in low-cut, narrow-waisted gowns of wool, the men in hooded tunics of the same material.

Even more telling is that while the starving settlers were slaughtering their remaining cattle and dogs, ringed seals swam in the nearby fjords, just under the ice. Yet researchers have not found a single harpoon from excavations of fifty Norse farms, despite the fact that the Inuit were experts at stalking and dispatching seals at their breathing holes. The same is true of the spiral-tusked narwhal and other species of whale that had to be taken by boat, the harpoon lines buoyed by air bladders made of sealskin. Having no wood with which to build ships in the European style, the Greenlanders refused to adapt to the skin craft of the Skrälings, choosing starvation to preserve their ethnic purity.

~

A coda to this melancholy fugue is now in the advanced stage of composition. Deep inside the insulated walls of the National Ice Core Laboratory in Denver, Colorado, are thousands of meter-long tubes arranged like so many library books on shelves. Within these silver, plastic-lined canisters are ice cores collected by scientists in Greenland and Antarctica. To guard against degradation, the temperature is kept at –36°C, requiring employees and visiting scholars to cover every part of their bodies except their eyes. Frozen in this artificial winter is a record of Earth's climatic history spanning the past 250,000 years.

Like the pages from a book, successive layers of frozen atmosphere from each Greenland summer and winter contain bubbles of gas, chemicals, and flecks of dust that made their way to the continent on air currents from the rest of the Northern Hemisphere. Trapped in falling snowflakes, these minute specimens settled on the surface, where they were quickly entombed in ice, one year's deposit covered by the next, much like layers of sediment. When drilled and subjected to sophisticated laboratory analysis, the ice cores become the climatologist's equivalent of the Rosetta stone, revealing secrets about violent storms, volcanic eruptions, precipitation, winds, and sudden shifts in temperature. Someone wishing to know the content of the air breathed by Socrates, Alexander the Great, or Augustus Caesar has only to analyze these ancient air bubbles.

The cores were drilled by two teams of researchers: the U.S. Greenland Ice-Sheet Project 2 (GISP2), which completed its work in 1993, and the Greenland Ice-Core Project (GRIP), a European effort that was concluded a year later, twenty miles away. The oxygen isotopes gathered from these frozen thermometers confirm both the archaeological evidence and the enamel record contained in the ground-up teeth of malnourished Norse children. Fourteenth-century Greenland experienced a clustering of chilly periods, punctuated by shorter and cooler than normal summers that led to starvation in the Western Settlement.

Meanwhile, the more heavily populated Eastern Settlement hung on, but barely. The colony's last bishop arrived in 1377, promptly died in 1378, and was never replaced. After

1408 regular contact with the colony was lost, although an occasional ship put in to trade or was brought involuntarily by storm. At some point during the fifteenth century, the last bodies, still wearing European garb, were buried in the graveyard, where they would be preserved by even harsher weather to come. "The darkness," as historian Peter Brent wrote, "was coming down, a night that would take them all at last."

In 1540 a German merchant ship, driven off course by a raging gale, made for a Greenland fjord. In outline above the passing shore were half-ruined stone buildings, like those the captain had seen in Iceland. Soon after landing, the crew came across the curled and frozen body of a man, his features masked by a woolen hood. Beside him lay a curved knife, its blade worn thin by years of repeated sharpening. How long he had been dead no one could say, but the spectral aura of silence and desolation told the visitors that before them lay Greenland's last Viking.

~

One of the more revealing findings of the scientists who study the Greenland ice cores is that for the past 8,000 years—virtually the entire Neolithic period—civilization has experienced an unusually warm and stable climate. By contrast, the 70,000 years before the domestication of plants and animals were climatologically chaotic, with much greater and more frequent temperature fluctuations than researchers, who were educated within a Darwinian time frame, could have imagined. At the end of the last ice age 12,000 years ago, a period known as the Younger Dryas, major temperature swings occurred as often as every five to twenty years. Temperatures rose by an astonishing ten degrees Celsius within the life span of a Paleolithic hunter, and some scientists now think that even that figure is too low by half.

It requires little imagination to realize that this sobering finding could signal major trouble ahead, especially if, as some climatologists conjecture, Earth's temperatures are due for another roller-coaster ride. Even if they are not, will the rise in global temperatures by about one degree Fahrenheit over the last century, attributed by many to the burning of fossil fuels, trigger harsh climatological changes fraught with dire consequences? Unimportant as the greenhouse effect may seem

compared with the natural order of things, it is well to remember that very small changes in temperature have literally
doomed civilizations. Nor should one be too easily comforted
by the fact that these events unfolded long ago and far away.
The fate of the Anasazi and the Vikings in Greenland serve as
analogues for one of the more harrowing climate shifts in living memory.

~

The prolonged cycle of drought and cool weather that forced
the Anasazi to abandon their ancestral lands more or less continued for another 600 years. Then, during the latter half of
the nineteenth century, at the very moment the secret of their
existence was being discovered, the climate changed once
more. The rains returned, and the temperatures moderated;
everywhere west of the Mississippi seemed to have turned
green, especially the southern region of the Great Plains that
includes parts of New Mexico, Colorado, Kansas, Nebraska,
Texas, and Oklahoma.

Farmers and ranchers poured into the largely virgin territory, whose burgeoning economy was built on cattle and
wheat. The coming of the iron-wheeled tractor during the late
1920s saw millions of acres of grassland succumb to the plow.
With wheat selling for a record dollar a bushel, more land and
more tractors were purchased as yields shot up from twenty to
fifty bushels an acre, a great golden tide that flooded every
grain bin and elevator for hundreds of miles around.

But the first of several devastating blows was about to fall.
Wall Street was shaken to its foundations in 1929, and by 1932
the bottom had dropped out of the agricultural market: Oats
were selling for ten cents a bushel, corn and barley for three
pennies more; wheat topped the grains at twenty-five cents,
one quarter of its former worth. Then nature entered into a
vicious conspiracy with the economy.

Annual rainfall began to decline in 1931, producing what
those who lived through it called "drouth," as if their tongues
had suddenly turned dry from memory. In Oklahoma, where
rainfall had been averaging seventeen inches a year, less than
twelve inches fell in 1932, fifteen in 1934, and barely nine in
1935. Conditions were no better in Nebraska, where in 1934 a
meager fourteen inches was the lowest amount of rainfall

since the Civil War. Nor was the drought confined to the Great Plains; during the 1930s, twenty states set records for dryness, many of which are still unbroken.

"Short crops" were soon replaced by "very short crops," which dwindled to "nothing." At the urging of the federal government, cattle were returned to the withering grasslands in numbers not seen since before World War I. While poorly designed plows churned the sprawling fields into fine powder, the hooves of the free-ranging beasts pulverized the thin, moisture-starved topsoil. Too late did Washington discover its mistake and initiate a program of buying and slaughtering cattle to remove them from the range. Dispatched by bullets where they stood, the slain beasts were left to decompose beneath a bloodshot Sun.

On a quiet Sunday afternoon in April of 1934, the prairie winds suddenly came rolling down the Plains from the north like a great tidal wave, preceded by darting flocks of panicked birds. In an instant day became midnight as livestock, machinery, and whole farms were obliterated by the enveloping dust. Ignition systems on cars shorted out; windmills, burnished by swirling topsoil, became electrically charged; people were instantly disoriented and lost their way only a few paces from home, some choking to death, reason enough to call the worst day of a terrible decade Black Sunday.

The dust storms kept on coming at a rate of up to seventy a year. In an attempt at classification, the Kansas Academy of Science labeled them "rectilinear," "rotational," and "ebullitional," then further divided them into "species" bearing such colloquial names as "sand blows" and "funnel storms." Dynamics aside, the winds deposited their burden in multicolored drifts against sheds and fencerows, red soil from Oklahoma, yellow from Kansas, other hues from other states. Abandoned cattle walked knee-deep across these new landmasses, only to find that they were bridges to nowhere. Thirst and starvation waited on both sides of the fence and sometimes claimed their victims halfway, front quarters and hind neatly divided by barbed wire. Only the coyote and the buzzard made out.

In a Dust Bowl dream of deliverance, the Dekalb Company, a struggling supplier of hybrid seed, grafted the wings of

Mercury onto its emblem of a golden ear of corn, but to no avail. Like their crops, people became straws in the wind. At sheriffs' auctions cattle, horses, machinery, and household possessions went on the block to satisfy mortgages held by banks that were themselves going under. By the end of the 1930s, 350,000 "Okies" had abandoned their homes and joined the nationwide migratory flow.

Like leaves in windrows, the dispossessed gathered beside the tracks of the Union Pacific, the Santa Fe, the Katy, the Rock Island. By night their fires blinked like the bivouacs of great armies on the outskirts of cities under siege. Begging, occasionally working, their way across the country, they cooked their meager fare in blackened tin cans, then slept with one eye open, wrapped in old newspapers in nameless hobo camps that grew more rapidly than the towns of the Alaskan gold rush. They fought with bare knuckles, straight razors, and switchblade knives, had intercourse under dirty blankets in the dark corners of crowded freight cars, converged into pairs and small groups, then dispersed again, only to repeat the ritual down the line. Most were eager to reach the next town but tired of it within hours and knew contentment only when the steel wheels were clicking beneath them, the half-buried fence posts hypnotically slipping by.

Those with the means loaded up their tread-bare jalopies and turned west, toward the setting Sun and the green promise of California—"the big rock candy mountain" of popular song. But the hand-painted signs along the roadside were ominous: Jobless Men Keep Going. We Can't Take Care of Our Own. On reaching the California line, they were rudely stripped of their last dreams of nirvana by jackbooted state troopers and local deputies wielding clubs and brass knuckles. Scum Not Wanted! read the placards. Okies Go Home!

Back in Oklahoma, half of the state was on relief, the days and nights spent battling the eternal dust. Dust in the beds, dust in the flour bins, dust in the cupboards and on the windows and walls, in the eyes and ears and nostrils and throat. Wet blankets hung in the windows; table manners were forgotten as people ate standing up out of pans on the stove because the dishes were coated with grit. Guide wires reaching from the house to the barn allowed families to venture forth to milk a remaining cow or two. At night they went to bed with dampened cloths over their faces, only to be greeted the next morning by ghostly outlines of their bodies on the dingy sheets and pillows. Yet it was the dust nobody saw that could be the most frightening. Slowly, silently it sifted through the siding and piled up in the attic until the joists gave way and the ceiling suddenly opened, releasing an avalanche of dark powder.

Ironically, the relentless wind scoured its way down into a long-forgotten past. Collectors of arrowheads and other artifacts prospered, little knowing that the Anasazi had been displaced by an earlier climate change with no less devastating consequences. Unlike the wind-driven Okies, the ancient residents of the Colorado and New Mexico plateaus had somewhere else to go. Scattering in several directions before the Great Drought, they became the Hopi, the Zuni, and the Pueblos of the northern Rio Grande, the spirits of their new kachina religion displacing the withered gods of old.

~

In 1937, during the height of the Dust Bowl and the Great Depression, Glen Thomas Trewartha, a forty-one-year-old associate professor of geography at the University of Wisconsin, published his well-regarded textbook *An Introduction to Weather and Climate.* In the first chapter, "Air and Temperature," the author describes how short-wave solar energy, absorbed at Earth's surface, is transformed into heat. The planet in turn becomes a radiating body by releasing much of the incoming solar energy back into the atmosphere as long-wave (infrared) radiation. Water vapor, carbon dioxide, and other less plentiful atmospheric gases absorb this energy by acting as an "insulating blanket." But Trewartha preferred to call it something else. "Obviously the effect of the atmosphere is analogous to that of a pane of glass, . . . thus maintaining surface temperatures considerably higher than they otherwise would be. This is the so-called *greenhouse effect* of the earth's atmosphere." That the middle-aged professor chose to italicize and index this phenomenon is explained by the fact that he seems to have been the first person to call it by its now familiar name, at least in print.

If Trewartha was suspicious that the record heat and

drought of the 1930s was caused by anything other than natural meteorological processes, he gave no hint of it. Not so the
self-anointed, who experienced terrifying visions and spoke in
tongues foretelling the end of the world. Others, skeptical that
divine retribution was imminent, put forth less dramatic theories. Perhaps it had something to do with that newfangled
plow made of disks strung out on a beam. After all, it barely
scratched the earth's surface, turning it into fine powder just
waiting for a big blow. Or maybe it was the tens of thousands
of cattle, which, unlike the buffalo, needed water on a daily
basis and stirred up the parched land like a moving churn.

Yet even if, as some experts now believe, the Dust Bowl
could have been averted by proper range management and
better-informed agricultural practices, how was one to explain
the dire climatic changes bedeviling the rest of the country?

In New York City the recurrent summer heat resulted in a
death watch reminiscent of London's final battle with the
Black Death in the mid-1660s, albeit on a lesser scale. On July
12, 1936, the *New York Times* printed a three-column list of casualties beneath the somber headline "Deaths and Prostra-
tions," not unlike the weekly Bills of Mortality issued by
English officials of the seventeenth century. The youngest to
die was one Albert Casey, a six-month-old infant; the eldest, a
widow named Emma Zimmermann, who had celebrated her
hundredth birthday in February. A few made it to the hospital,
but most succumbed in their homes or at work. Armus Goldie
of Orchard Street collapsed on the roof of his house, falling to
his death in the courtyard below; John Wash, an employee of
John A. Roebling's Sons, the cable manufacturer whose audacious founder had undertaken construction of the Brooklyn
Bridge, was overcome at the factory and died. Among the
many others were four men and women simply listed as
"unidentified," street people of another time.

The same thing happened the following year to the very
day. On July 12, 1937, a Monday, every major beach in the
New York City area broke all attendance records. Coney Island
and the Rockaways each tallied more than 1 million visitors,
while Jones Beach recorded 144,000. When the Sun set, thousands chose to stay, preferring to spend the night on the cool
sand rather than return to a brick-and-concrete city on slow

bake. As the number of deaths mounted, the size of the newsprint in the casualty lists shrunk inversely. Across the nation 295 had succumbed in the latest heat wave, with no relief in sight. The widespread gloom deepened when word flashed across the wires that the composer George Gershwin had died in Hollywood at thirty-eight, after surgery for a brain tumor. During the next few evenings radio stations swaddled the country in *Rhapsody in Blue* and "Summertime," the deeply wounded economy and searing temperatures undermining Bess's refrain that "the livin' is easy."

11

THRESHOLD

~

My candle burns at both ends;
It will not last the night;
But, ah, my foes, and, oh, my friends—
It gives a lovely light.

—Edna St. Vincent Millay, *A Few Figs from Thistles*

When the astronomer Edwin Powell Hubble first arrived in California in September of 1919, after completing his tour of duty in World War I, Pasadena was little more than a quiet village set among vineyards and orange groves intersected by rambling dirt roads. Overhead stretched pristine skies, star-spangled and black as velvet after sunset, the lifeblood of the astronomer. No smokestacks or exhaust or bright city lights obscured the magnificent images streaming earthward time out of mind.

All that was about to change. At the end of Colorado Street stood the elevated remnants of an abandoned dream, the never completed cycleway that would have allowed locals to pay a small toll for the privilege of biking to Los Angeles undeterred by the weather, rutted roads, and horse droppings. Construction had ceased when it became clear to investors that the automobile had rendered the structure obsolete before they could collect their first dime.

Christmas Eve found Hubble atop Mount Wilson detached from this small planet circling its middling star, his fingers and thumbs poised on the buttons of the giant one-hundred-inch Hooker telescope, the greatest gift any astronomer could desire. During the coming decades, his discoveries would revolutionize humanity's conception of the cosmos, for it was Hubble who provided conclusive proof that the Milky Way is not the universe entire but merely one of untold billions of galaxies, and that these massive congregations of stars are hurtling away from each other, the result of a primordial explosion later dubbed the "Big Bang."

After he had become well known, Hubble would walk visiting astronomers and Hollywood actors to the brink of the mountain, where they gazed silently on the beauty in the distance. Below them stretched a Persian carpet woven of alternating shades of manufactured glitter, the nearer lights twinkling like fallen stars. Just above were long streaks of luminous fog, reminiscent of the great band of the Milky Way itself. Far off, blurred by haze and summer heat, lay downtown Los Angeles, which Hubble referred to as the dense nucleus of the "Los Angeles nebula."

Yet as a professional astronomer he despised the illuminated city. The suburbs below sprawled across nearly the whole of the mountain-ringed plain, their lights a shimmering lava flow in reverse, creeping up the heights to fog the sensitive photographic plates where the "seeing" had once been the best on the planet. Worse still, a strange acrid haze, known locally as "smog," was slowly enveloping the 5,714-foot peak, dimming the far-off galaxies and with them the hopes of anxious astronomers.

A smog committee was formed and Hubble became a member, although he knew that their efforts would almost certainly fail. Instead, he was pinning his dreams on Mount Palomar Observatory, then under construction, northeast of San Diego, where only the campfires of the nut-gathering Palos Indians flickered in the distance. During one of the Hubbles' many voyages to Europe, Edwin's wife, Grace, gazed out of their porthole and wrote wistfully in her diary that no matter what horrors human beings were inflicting on the fair face of the land, they could do nothing to harm the sea.

~

The recipe for smog is a fairly simple one: Begin with a large measure of oxygen and carbon monoxide, add a dash of nitric oxide together with some sulfur compounds, mix with sunshine to promote a photochemical reaction, and place in an inescapable container such as the Los Angeles basin.

Overlying the city and the rest of the planet is the lowest part of the atmosphere, a region that extends from Earth's surface to about seven miles (twelve kilometers) above sea level. Known as the troposphere, it contains roughly 80 percent of the mass of the atmosphere and is the home of all the weather on the planet. Under normal conditions the troposphere is warmest at the planet's surface, heated from below by the infrared radiation of the greenhouse effect. In general, temperature decreases with height at the rate of about 6.5 degrees Celsius per kilometer of altitude, or 3.6 degrees Fahrenheit for every 1,000 feet of elevation. Unfortunately for Los Angeles, the normal rules don't apply. On a typical summer afternoon the temperature above the city rises with altitude, creating the condition known as an inversion.

The factors that combine to produce this phenomenon are both geologic and climatological in origin. Heated air at the surface normally forms invisible convection currents that rise like steam from a boiling kettle. However, in the case of Los Angeles, which is walled in by mountains on the east, the heated air cannot escape. Instead, the sea breeze blowing inland at low levels pushes the warm polluted air against the mountains, holding it in place. Because it is cooler than the atmosphere above it, the trapped air is inverted, the opposite of the normal configuration. In the evening the pollution decreases as harried commuters reach home and the breeze reverses itself, blowing back out to sea, while the catalytic powers of the Sun are neutralized. With the dawn of another workday the grim cycle begins anew as millions of automobiles take to the freeways.

More than any other factor, the perfection of the internal combustion engine early in the twentieth century made petroleum one of the most important energy sources in the world. Following the example of Colonel Edwin Drake, oil drillers launched a boom in fossil fuel production unparalleled in human history. California wrested the lead from Pennsylvania in 1903, only to be surpassed by Oklahoma in 1907, with which it vied for supremacy until 1928, when Texas outproduced both. Indeed, the so-called golden age of American oil discovery began with major finds in 1930 in East Texas. By 1918 Mexico had become the second-largest producer in the world, though it quickly faded in the wake of political upheaval and depression. Then, in the 1930s, the Middle East suddenly entered the picture, as huge petroleum fields were discovered in Iran, Iraq, and Saudi Arabia, the pipelines from which carried the raw fuel to the Mediterranean Coast for loading aboard the tankers of the Western powers.

Behind it all loomed the gray eminence of mechanical technology. Born during the Civil War in Dearborn, Michigan, Henry Ford had demonstrated an engineer's aptitude at an early age and left his father's farm to work as an apprentice in a Detroit machine shop. He soon came home, where he began to experiment with power-driven vehicles. Armed with new knowledge and insight, he returned to Detroit and was hired as a machinist and engineer with the Edison Illumination Company. Ford continued working on his first automobile in his spare time, conducting his initial test-drive in 1892. He resigned from Edison five years later and launched his first manufacturing venture, the Detroit Automobile Company. A subsequent falling-out with his associates led him to found the Ford Motor Car Company in 1903.

At this point automobiles were still basically handmade from parts that had only short production runs. This made them expensive to purchase as well as costly to repair, effectively limiting the market to the well-to-do, who used them mostly for recreation. In the beginning Ford utilized the methods of his competitors but soon despaired of getting ahead in the overcrowded market. Gambling everything he owned, he decided to revolutionize operations: first, by gaining control of raw materials and the means of distribution; second, by adapting the conveyor belt and assembly line to mass production. In his mind's eye Ford pictured an inexpensive car, painted black, and constructed of standardized parts that would be easy to install.

In 1908 the first of 15 million Model T's rolled off the assembly line. The "Tin Lizzie," as it was affectionately known, featured a rust-resistant brass radiator, a rugged four-cylinder engine cast in one piece, and a detachable cylinder head for easy repairs. By 1913 it accounted for 40 percent of U.S. pro-

Henry Ford and the Model T
(From the collections of the Henry Ford Museum & Greenfield Village)

duction; by 1920 half the cars on the planet were Model T's. Long before it was finally replaced by the redesigned Model A in 1928, the Tin Lizzie had fulfilled Ford's dream of becoming "the family horse." Along the way, he had created a sensation by doubling the standard daily wage—to five dollars for an eight-hour day—a stroke of genius that dramatically reduced labor turnover, enabling the manufacturer to amass a personal fortune estimated at well over half a billion dollars.

Ford invested some of it in Greenfield Village, a reproduction of an early American town located in his native Dearborn. There, in the early evening, he occasionally walked the quiet streets in the company of a few friends, waxing nostalgic on a time and place obliterated by his own invention.

In 1929 the number of automobiles in the United States reached 26.7 million, a swarm so great that the total mileage of surfaced roads overtook that of the railroads in 1915 and continued expanding rapidly, reaching 500,000 miles in 1925, then doubling to an even million a decade later. But by now Ford's preeminence as the nation's largest producer of cars was lost to his competitors, largely because of his hesitancy to embrace the recent innovation of introducing a new model each year, a marketing ploy that delighted restless consumers.

The lead belonged to General Motors, which was bent on exploiting its advantage by eliminating as many cheap and efficient alternatives to the automobile as possible. Starting in the 1920s, GM formed a series of holding companies that began purchasing dozens of electric surface rail transit systems throughout the country, including those in smog-free southern California, which had the largest interurban network in the land. The most successful of these was the Pacific Electric Railway Company, popularly known to its patrons as the "Red Cars." At its peak, the company owned more than 1,100 miles of track and annually logged millions of riders over a three-county area.

General Motors became interested in 1936 and formed a holding company called National City Lines. The plan was to displace rail transportation with GM-manufactured diesel buses, then displace the buses with automobiles. Two of GM's major suppliers, Standard Oil of California and Firestone Tire, joined the conspiracy. Having already been successful in buying and scrapping electronic rail systems in Fresno, San Jose, and Stockton, National City began to acquire and dismantle part of Pacific Electric, ultimately motorizing downtown Los Angeles. Patrons of the quiet, high-speed electrical rail service were not used to the noisy, foul-smelling buses and abandoned them by the thousands. The sales of large, fuel-hungry automobiles skyrocketed, causing Ford and Chrysler to follow GM's example by eliminating other carriers. In the mid-1940s, GM and its allies severed Los Angeles's last regional rail links. Where 3,000 trolley cars had once been in operation, not a single one remained. The same strategy played out in another forty-five cities, where more than a hundred electric transit systems were replaced by GM buses. Though no one was aware of it as yet, social and environmental policy of the United States had been irrevocably altered at corporate hands.

~

As the trolleys disappeared and the Okies took to their overloaded Model-T's and headed west to the land of the giant telescopes, hardly anyone was pondering the question of whether humans were unwittingly influencing the climate through rapid industrialization. One of the few exceptions was an English coal engineer named George S. Callendar, who held the position of steam technologist to the British Electrical and Allied Industries Research Association. In February of 1938, Callendar published the first article on the greenhouse effect in decades, titled "The Artificial Production of Carbon Dioxide and Its Influence on Temperature."

As best Callendar could determine, humans had added to the air since the 1880s about 150,000 million tons of carbon dioxide, three quarters of which was still suspended in the atmosphere. "Few," he wrote, "of those familiar with the natural heat exchanges of the atmosphere, which go into the making of our climates and weather, would be prepared to admit that the activities of man could have any influence upon phenomena of so vast a scale." Yet "I hope to show that such influence is not only possible, but is actually occurring at the present time."

Utilizing data collected from 200 weather stations around the world between 1880 and 1934, Callendar determined that Earth's temperature had been on the rise during the previous fifty years. The most significant increases had occurred in the Northern Hemisphere, where industrial production was concentrated. Two periods in particular stood out: 1920–29 and the depression years 1930–34 (which experienced the greatest temperature rise of all). In sum, the planet had undergone warming on the order of one degree Fahrenheit, a figure that Callendar predicted would double in the next half century, since the end of industrial output was nowhere in sight and so many lingering particulates were already in the atmosphere.

Anticipating his critics, the engineer had wisely factored in the heat retention that occurs in large cities, where buildings are tightly clustered, tending to skew temperature measurements, a phenomenon known as the urban heat island effect. He also noted that while the oceans of the world may well serve as vast sinks for the absorption of excess carbon dioxide, this process mostly takes place along the coastline, where the shallow water is constantly exposed to the air via wave action. However, this contact area accounts for only 1.3 percent of the oceans' volume. Below a depth of 200 meters, absorption occurs at a snail's pace, if at all. While much was yet to be learned about the vertical circulation of the oceans, "several factors point to an equilibrium time, in which the whole sea volume is exposed to the atmosphere, of between two and five thousand years." Simply put, the seas could not be counted on to neutralize the rising quantities of carbon dioxide for at least another hundred generations.

Prescient though Callendar was, he did not look upon his conclusions as the least bit sobering. Instead, he cast his lot with an optimistic Svante Arrhenius. "In conclusion it may be said that the combustion of fossil fuel, whether it be peat from the surface or oil from 10,000 feet below, is likely to prove beneficial to mankind in several ways, besides the provision of heat and power." Small increases in mean temperature would be a boon to agriculture at the northern margins of cultivation; plant growth would increase significantly with the rise of carbon dioxide and, most important, the deadly glaciers of the Pleistocene would return no more.

~

Other than some comments in one of his popular books, Arrhenius, who had been appointed director of Stockholm's Nobel Institute, wrote nothing more about the greenhouse effect after his groundbreaking article appeared in 1896. The supposedly benign influence of atmospheric warming was relegated to some distant future, of no pressing concern either to the scientist or to the layperson. Still, he could not rest easy. Arrhenius had become preoccupied with a related and, to him, truly urgent problem. As he pondered the rapid increase in the consumption of fossil fuels, he came to believe that the world was faced with a crisis of catastrophic potential, the nature of which he addressed in the last of his popular books, *Chemistry in Modern Life.*

Global use of coal had risen from an estimated 800 million tons in 1903 to 1,305 million tons in 1920. Based on the current level of consumption, projections said that Great Britain would exhaust its coal reserves in four and a half centuries, while Germany's supply would be used up in 1,000 years. Even the United States, though much richer in carbon deposits, could expect its finite supply of coal to give out in another 4,000 years. In all probability, however, consumption would continue unchecked and undermine this forecast. The last of England's mines could close in as few as 50 years, followed by those of the United States a century after that.

More sobering still was the gluttonous consumption of oil, spurred by the invention of the automobile and its mass production in the factories of Henry Ford and others. Some of the earliest petroleum fields, including those in western Pennsylvania, had already been severely depleted, and known reserves were fast dwindling. The United States would likely pump its last barrel in 1935, a mere twenty years in the future. Even if demand were to remain steady over the coming decades, data collected in 1921 indicated that the shutdown of the final well would occur in 1953, at the latest.

Having lived against a backdrop of rampant imperialism and world war, Arrhenius envisioned a return to international chaos: "Concern about our raw materials casts its dark shadow over mankind. Those states which lack [them] throw lustful glances at neighbors, which happen to have more than they use. Still more tempting is the desire for gain from lands on the other side of the seas, inhabited by uncivilized natives, with interest unawakened to guardianship." The industrial world had given rise to a new kind of international warrior, the "conquistador of waste."

And what of future generations? "Like insane wastrels, we spend that which we received in legacy from our fathers. Our descendants surely will censor us for having squandered their just birthright."

The solutions, like the warnings of this prophet of the energy crisis, strike a now familiar chord. "Statesmen can plead no excuse for letting development go on to a point where mankind will run the danger of the end of [natural resources] in a few hundred years." Invoking the Chemist's Commandment—"Thou Shalt Not Waste!"—Arrhenius urges that legislation be enacted, aimed at both reducing consumption and promoting conservation. His political hero is the voluble Teddy Roosevelt, a recipient of the Nobel peace prize, who, if given half a chance, would have killed two of every species that survived Noah's ark. At the same time, it was Roosevelt who added millions of acres of land to public ownership, and between 1908 and 1909 had tried without success to rally the nations of the world and the U.S. Congress to the conservationist cause. Then came World War I and everything else was pushed aside.

To conserve coal, a half ton of which is burned in transporting the other half ton to market, Arrhenius advocates the building of power plants in close proximity to the mines. All lighting with petroleum products should be replaced with more efficient electric lamps, while aluminum, the virtually limitless metal of the future, should be substituted for iron, whose ore reserves are finite and rapidly dwindling. Engineers must design more efficient internal combustion engines capable of running on alternative fuels such as alcohol, and new research into battery power should be undertaken. Wind motors and solar engines hold great promise and would reduce the level of carbon dioxide emissions. Forests must be planted, dams raised, and atomic energy explored, although the prospect of taming the atom seemed exceedingly remote in the 1920s.

For all his concern, which was surely heartfelt, Arrhenius was infused with the easy confidence of Faustian man: "Doubtless humanity will succeed eventually in solving this problem. . . . Herein lies our hope for the future. Priceless is that forethought which has lifted mankind from the wild beast to the high standpoint of civilized humanity." Somehow, it never occurred to him that the headlong consumption of fossil fuels, coupled with the massive release of their invisible gases into the atmosphere, might be tantamount to burning the candle at both ends.

12

A TAP ON THE SHOULDER

~

While I nodded, nearly napping, suddenly
there came a tapping,
As of someone gently rapping, rapping
at my chamber door.

—Edgar Allan Poe, "The Raven"

While George Callendar sorted and totaled some fifty years' worth of temperature data, a bespectacled chemist living in Dayton, Ohio, was engaged in what at the time seemed a noble quest. Thomas Midgley Jr. had joined the staff of the General Motors Research Corporation, otherwise known as Delco, during World War I and quickly fulfilled his promise by developing lead tetraethyl, an antiknock agent that when combined with gasoline raised compression ratios in airplane engines, increasing power while reducing wear. The effervescent "Midge" became an instant hero and GM's profits soared when tetraethyl lead was added to the gasoline used in automobiles following the war.

Decades would pass before anyone realized that this major technological breakthrough had its dark and insidious side. Lead not only made its way into fuel but was added to paints and many other products harmful to the environment. Only with the advent of the catalytic converter was leaded gasoline banned in the United States, Canada, Japan, and much of Europe, while the battle to rid homes and apartments of toxic paint continues at untold cost.

With his star in the ascendant, Midgley was given another research challenge at the onset of the depression. Dr. Lester S.

Keilholtz, chief engineer of the Frigidaire Division of General Motors, came to Dayton to discuss the growing demand for artificial refrigeration, and none too soon given the record temperatures in store. Frigidaire used sulfur dioxide as its cooling agent, the poisonous nature of which had recently been singled out by the prestigious American Medical Association, the kind of adverse publicity that causes corporate executives to panic. The alternative agent, methyl chloride, was just as toxic, while ammonia, the major industrial refrigerant, had recently killed several people in a Cleveland hospital following an accidental spill. To make matters worse, state and municipal agencies were thinking of requiring manufacturers to add a number of ill-smelling chemicals to their coolants as a means of warning consumers of potentially hazardous leaks. Keilholtz didn't mince words in his discussions with Midgley: "We must have a nontoxic, nonflammable, cheap refrigerating agent, quick! What can you do about it?"

In frantic cadence imitative of his labors, Midgley described what followed: "Plotting boiling points, hunting toxicity data, corrections; slide rules and log paper, eraser dirt and pencil shavings, all the rest of the paraphernalia that take the place of tea leaves and crystal spheres in the life of the scientific clairvoyant were brought into play."

Having successfully used the periodic table to narrow the possibilities in his search for an antiknock agent, Midgley returned to the paper-and-pencil approach, which quickly bore fruit. The atomic numbers told him that the unlikely combination of chlorine, fluorine, and methane would do the trick, but certain of the chemicals were in such short supply that the shelves of laboratories across the country had to be scoured for their tiny stocks. A few months later the inventor took the train to Atlanta and appeared before his fellow members of the American Chemical Association. In theatrical fashion, Midgley breathed in a gulp of his new chemical, then exhaled it into a rubber tube that ran under a bell jar. The burning candle inside was quickly extinguished, proof that this new compound—dubbed dichlorodifluoromethane—was neither poisonous nor flammable.

Other artificial chemicals, whose names would one day bedevil spelling bee contestants and surface on rigged quiz shows, soon followed. Midgley called them chlorofluorocarbons, or CFCs, each cobbled together in the laboratory by linking atoms of chlorine, fluorine, and carbon. Yet, chemically speaking, they were as stable as stones and cooled with the efficiency of a Greenland winter.

Midgley's hybrid offspring were soon being manufactured and marketed under the name Freons by E. I. Du Pont de Nemours and Company. The air-conditioning business in the United States expanded by sixteen times between 1930, the first year of production, and 1935. During the 1960s production rates rose at a robust 20 percent a year, spurred by the use of CFCs in automobile air-conditioning systems and as spray-can propellants and foam-blowing agents. With the disappearance of the vacuum tube, they became indispensable in the cleaning of circuit boards that go into television sets and computers.

Although Midgley was the holder of 117 U.S. patents, his reputation as a "genius for benign invention" was to be sullied a second time by another series of unforeseen developments. Year after year, decade after decade, the nearly impervious CFCs drifted upward through the troposphere and into the stratosphere beyond, where they accumulated by the millions of tons and attacked the ozone, the protective layer that shields Earth's creatures from the damaging ultraviolet radiation emitted by the Sun. Moreover, CFCs themselves are highly efficient greenhouse gases that endure upwards of a century or more.

Despite an occasional warning from a small circle of wary scientists, the first important paper suggesting that CFCs could destroy large amounts of stratospheric ozone would not be published until 1974, fully thirty years after Thomas Midgley's tragic demise. In 1940 the gregarious chemist contracted polio, which left him crippled for life. He battled back by designing an ingenious harness-and-pulley system to help get himself out of bed in the morning. But with both legs paralyzed and his general health deteriorating, he grew despondent. On a November day in 1944, while he was alone in his room, Midgley intentionally strangled himself in the harness rigging of his own invention, thus sparing him the additional pain of learning that the great good he had done also contained the seeds of great harm.

~

Callendar's 1938 paper had launched a pointed albeit brief debate among his colleagues in the Royal Meteorological Society. While they praised him for his "courage and perseverance," they were understandably skeptical. It was argued that the gradual rise in temperatures during the previous thirty years was no more striking than changes that had occurred in the eighteenth century, when the issue of human intervention was hardly in play. Moreover, the temperature increase was about ten times as great in the arctic regions as in the middle and low latitudes, a differential that was difficult to attribute to a rise in the levels of carbon dioxide. Indeed, was the carbon dioxide in the air really increasing? Then there was the question of whether Callendar's calculations were wholly trustworthy. It was a nice try nonetheless, and they welcomed Old George's paper as a "valuable contribution to the problem of climatic changes."

World war followed on the heels of global depression, and no one—certainly not scientists whose skills were desperately needed by their governments—was paying any attention to the so-called greenhouse effect. No one, that is, but a bulldoggish Callendar, who had taken on the arduous task of scouring the literature of many countries for numerical data on the rise of carbon dioxide in the atmosphere, harmless though he still believed this process to be.

~

It was not without a degree of national pride that Callendar and his fellow Londoners called them "pea-soupers"—the great fogs that buried their city under a stifling dome of yellow and brown haze, or what Lord Byron had called "a dun colored cupola." Yet not until the final decades of the Victorian era did this stagnant shroud begin to return every autumn, with the regularity of some migratory ghost. The Yankee poet James Russell Lowell, who was serving as U.S. minister to England in October of 1883, wrote that "we are in the beginning of our foggy season, and today are having a yellow fog, and that always enlivens me, it has such a knack for transfiguring things." Lowell considered himself fortunate to be a member of "that exclusive class which can afford to wrap itself in a golden seclusion."

The American-born English poet T. S. Eliot also drew inspiration from the jaundiced miasma. In his early masterpiece

"The Love Song of J. Alfred Prufrock," Eliot writes of "the yellow fog that rubs its back upon the window panes." As a youth, he had likely read Sir Arthur Conan Doyle's "The Adventure of the Bruce-Partington Plans," wherein an uncharacteristically glum Watson peers into a four-day fog, pronouncing it "a greasy heavy brown swirl still drifting past us and condensing into oily drops upon the window panes."

It was behind just such an impenetrable barrier that London's greatest true-life mystery unfolded. On the night of August 7, 1888, the first of seven prostitutes was murdered in Whitechapel, her throat slashed and her body methodically—some said surgically—eviscerated. The last victim died the same horrible death in Spitalfields on November 10, after which the killer quite literally vanished into the fog, leaving befuddled authorities with only a number of taunting notes from a person who signed himself "Jack the Ripper." It so happened that all the murders took place in the East End, which not only had the city's largest concentration of streetwalkers but had the most factory smokestacks and domestic chimneys as well.

While films, plays, and more than a hundred books on the Ripper continue to fascinate, London's industrial equivalent of the unknown serial killer—though more deadly by a factor of thousands—has been largely forgotten, except by those with an interest in medical or environmental history.

In December of 1873, a toxic yellow fog gripped London for three days, claiming as many as 700 of its citizens, the majority of whom succumbed to respiratory distress. Another 700 or more died during a similar assault in January of 1880. As the mortality lists grew following the turn of the century, Dr. H. A. Des Voeux, a member of the Smoke Coal Abatement Society, arranged for a more thorough analysis of the air during one such golden onslaught. The results confirmed the doctor's suspicions that in addition to the usual concentration of water vapor and carbon dioxide, the polluted air contained deadly amounts of sulfur oxides, chlorine, ammonia, ash, and what was deemed "tarry matter," the oily substance that had formed droplets on Watson's and Prufrock's imaginary windowpanes. Once he was satisfied that Londoners were dealing with a phenomenon of another order, Des Voeux reported his findings to the Manchester Conference of the Smoke Abatement League of Great Britain in 1911. He proposed that in the future ordinary fog be distinguished from the toxic brew he dubbed "smog," for smoke-fog, a term that struck a responsive chord among the public.

Unlike the photochemical smog of the Los Angeles basin, which has nothing to do with either smoke or fog, Des Voeux's aptly named concoction was a by-product of the domestic and industrial burning of high-sulfur coal. The resulting smoke and sulfur dioxide were occasionally trapped for days between London's surrounding hills and a stagnant mass of warm air, rising some 3,000 feet above the ground. Not only did the East End produce the most smoke, it occupied the lowest-lying land, inhibiting the dispersal of pollutants, which even on the best of days reduced winter sunshine by 30 percent.

Memories of London's great "killer smogs" had all but faded by December 4, 1952, when the Thursday forecast in the *Times* called for sunshine, light winds, a high temperature of 38° to 43°F, and the probability of late-evening frost. But by midmorning the breeze had stilled and conditions began to deteriorate as the skies became gray and the air grew damper. The next morning, citizens awoke to a "fog" thicker than all but a few could remember, and it continued to build throughout the day. Those who ventured out-of-doors experienced an intense burning sensation in their eyes and lungs, while their clothing became sooted within minutes. Friday evening saw the number of hospital admissions for respiratory difficulties double, a figure that continued to rise at an unprecedented rate. With the weekend at hand, most stayed home and attempted to fight off their claustrophobia by adding more coal to the fire.

Pollution levels surpassed all records. At the National Gallery, the filters of the air-conditioning system were clogging at twenty-six times the usual rate and peaked at fifty-six times normal. Wembley Stadium saw its first football match postponed since its opening in 1923, while the opera's performance of *La Traviata* was halted at the end of the first act because the theater was full of smog. By Monday, the eighth, all bus, taxi, rail, river, and air services had ground to a halt, and abandoned automobiles littered the streets as visibility was constricted to a scant fifteen feet. Scotland Yard was running itself ragged in response to a flood of burglaries, assaults, and robberies; members of ambulance and fire crews had to walk in front of their vehicles to avoid wrecking. At Central Line Station, 3,000 people queued up to purchase tickets on the Underground, the one transportation system still operating. Then, for good measure, a sharp frost transformed roads and sidewalks into polished glass, adding broken bones and concussions to the casualty lists.

Emergency services were overwhelmed as the Great Smog reached its peak. Mortality rates shot up from 250 per day to more than 900. By the time it had ended five days after its onset, the blanket of pollution was credited with playing a part in the deaths of 4,000 Londoners, about four times more than had died in 1892, the year of the city's second-worst smog. Among the witnesses to this calamity was a stunned George Callendar, who was still gathering data on the increase of carbon dioxide in the atmosphere. By the time he was ready to publish again after this ominous tap on the shoulder, he would have nothing more to say about the forgiving nature of the atmosphere.

~

While Parliament began inching its way toward the passage of the Clean Air Act of 1956, the first of its kind, a newly minted Ph.D. in chemistry from Northwestern University was in search of his first job. Hoping to dodge what he called the "pitfalls of practical science," Charles Keeling shunned the path chosen by most of his young colleagues, who wound up "mak[ing] breakfast cereals crisper, gasoline more powerful, plastics cheaper, and antibiotics more expensive." So it was that in 1955 the scientist was camping with his wife and infant son at Tioga Pass in Yosemite National Park. Awakened by a rustle in the small hours, Keeling groped for his flashlight and was greeted by two big eyes. He could just make out the dim object between the intruder's clinched teeth, causing his heart to sink. The hungry mule deer promptly absconded into the woods with Keeling's notebook, the panicked scientist stumbling after it through the darkness.

Two years earlier, Keeling had attracted the attention of Harrison Brown, head of the new geochemistry program at the California Institute of Technology, who had offered him a temporary position. Soon after Keeling arrived in Pasadena, he brashly challenged a remark Brown made while the professor was engaged in shoptalk with his fellow geochemists. It was Brown's contention that the carbon dioxide in the lower atmosphere is in balance with the CO_2 in the oceans and seas. Keeling argued that the two might just as likely be out of balance, although he later admitted that it was only a hunch, with no scientific data to back it up. After gaining Brown's approval, he set himself the task of finding out.

The solution to the problem seemed straightforward enough. Keeling would simply measure the amount of CO_2 in a stream, and in the air above it, then determine whether the gas pressures were equal. However, he immediately hit a snag that convinced him that his proposed study had no practical significance, which made it that much more appealing to the intellectual rebel. According to the supply catalogs, not a single manufacturer produced an instrument capable of measuring CO_2 in parts per million. Yet to abandon his quest was unthinkable, leaving Keeling with no choice but to build his own device. He rummaged through the scientific literature and finally found a 1916 article describing the manometer, a Victorian-sounding name for an instrument capable of measuring small amounts of various gases. If the design was properly modified, it just might do.

A perfectionist by nature, Keeling tinkered away while Brown wrote an entire book, *The Challenge of Man's Future.* He inscribed a copy to Keeling as "a man who knows what he is doing." "That," Keeling later quipped, "was *before* I knew what I was doing!" In point of fact, he was designing his manometer to take measurements that were ten times more precise than "made any sense."

The moment of truth finally came in March of 1955, a year and a half after Brown's offhanded remark. Keeling pumped the air out of an empty glass sphere in his laboratory, sealed it, and then carried it up the steps to the roof of Caltech's Mudd Hall. He opened the stopcock while holding his breath, so as not to contaminate the inrushing air with his own CO_2. After resealing the vessel, he returned to his laboratory and isolated the CO_2 with liquid nitrogen. He next put the CO_2 into a chamber of his completed manometer and compressed the gas with a column of mercury. The air he had just collected contained 310 parts per million of CO_2, which he recorded in a green notebook purchased especially for the occasion.

Mania ensued. Keeling began collecting air samples every four hours, twenty-four hours a day, either sleeping on a cot in his lab or leaving a warm bed and a pregnant wife to return in the middle of the night. Their son was only a few months old when the three took to the open road, where Keeling filled his flasks with air from the Inyo Mountains, the Cascades, Olympic National Park, and Yosemite, scene of his encounter with the thieving deer. The green notebook contained all of Keeling's data, and he forgot about the black bear he had seen only hours before as he desperately scanned the snowy land- scape with a flashlight. A glint of color caught his eye, and he rushed forward to discover his notebook, its binding torn away and its pages indented by a large set of teeth. He repaired the damage with a few strips of tape and was back in business.

On returning to his lab, Keeling made a fascinating if inex- plicable discovery. No matter where he had collected his sam- ples, those gathered in midafternoon yielded 315 parts per million of CO_2, or very near it. He packed up more flasks and headed back into the mountains, this time alone. He reached a remote field station across from Mount Whitney during the onset of a winter gale. Refusing to retreat, the chemist held on for five miserable days at 12,000 feet, fighting the driving winds and horizontal snow as he filled another thirty spheres with what has been described as "Essence of Pacific Mountain Storm." With his latest treasure secured, he headed back down to civilization and his waiting manometer.

The results were the same as before: 315! 315! 315! The number became an idée fixe and seemed to possess mystical qualities, like some figure from the cabala or the prophecies of Nostradamus. Then it hit Keeling like a speeding truck. The air he had been collecting beyond the polluted environs of the city was a blend of all the air on Earth, the level of CO_2 repre- sentative of the planet as a whole. The magic number was 315 after all!

~

Meanwhile, George Callendar's perseverence had paid off. Employing the most reliable data gathered in Europe, North America, South America, and Australia between 1866 and

1956, he calculated the base value of atmospheric CO_2 in 1900 at 290 parts per million. While he realized that many of the sampling techniques were far from perfect, he believed that his figures were "reasonably accurate" because of the frequent measurement of CO_2 levels toward the end of the nineteenth century. He took the added precaution of excluding data that deviated 10 percent or more from the general average of the time and region, as well as that derived from air samples collected in cities, where CO_2 levels exceeded those of uncontaminated air by as much as 20 percent.

Based on the remaining data, Callendar constructed a graph whose upward curve paralleled the increased burning of fossil fuels through the mid–twentieth century. Average CO_2 levels had risen from 290 ppm in 1900 to 325 ppm in 1956. Although Callendar's estimate was some 10 ppm greater than the figure yielded by Keeling's painstaking measurements, the two were clearly on the same track. Whether, in time, the vast waters of the planet would act as a giant sink and absorb the excess CO_2, Callendar still could not say, "although one suspects," he wrote in the last sentence of his last article, "that the final answers to this problem lie in the chill darkness of the ocean abyss."

~

Keeling held an untenured position, and by his third year at Caltech his research was being paid for by a grant from the Los Angeles Air Pollution Foundation. As he later explained, "I had inadvertently demonstrated that carbon dioxide was a tracer of urban pollution." However, it was not a subject that he was interested in making his lifework. He much preferred to go right on compiling the most accurate measurements possible, to the millennium if need be.

His luck held. In 1956, a scant two weeks after his air-collecting foray on a storm-buffeted mountain, Keeling was on a plane en route to Washington, D.C., his fare paid for by the U.S. Weather Service. Scientists from seventy countries were about to launch in 1957 the first International Geophysical Year (IGY), during which Earth and its atmosphere would be probed and prodded with unprecedented intensity. Those involved in the meteorological research were interested in measuring CO_2 levels around the globe to see if Keeling's number 315 was as universal as he believed.

Among those present at the 8 A.M. meeting in Washington was Roger Revelle, a distinguished student of the seas and director of the Scripps Institution of Oceanography in La Jolla, California, just up the coast from San Diego. Unbeknownst to Keeling, Revelle had been collaborating with Dr. Hans Suess, a fellow oceanographer at Scripps, on an article scheduled for publication in the journal *Tellus*. They had been studying the level of CO_2 exchange between the atmosphere and the oceans, as well as the question of an increase of atmospheric CO_2 during the past several decades. Thus far, the two were siding with Callendar, but neither their data nor Callendar's painstaking archival research on rising CO_2 levels was conclusive. Revelle wanted someone to settle the issue during the upcoming IGY.

Keeling found himself in the hot seat. Exactly how, his interrogators wanted to know, would he confirm that 315 was the global touchstone?

Without flinching, he replied that there was only one sure way. A new generation of manometers, costing about $10,000 each, would have to be built and placed in remote locations ringing the planet. Once installed, they would monitor the air for CO_2 by means of an infrared beam. As the beam penetrates the air sample, sensors measure how much of it passes through. The more CO_2, the greater the blockage. The process would be continuous and automatic, nothing like Keeling's crude practice of filling flasks by hand at all hours of the night.

Keeling made an impression. A few weeks later, at Revelle's invitation, he cleaned out his Caltech office and headed down the coast to Scripps. With ample funds at his disposal, he was soon engrossed in building a manometer that he promised Revelle would improve his measurements by a factor of ten.

In 1957, while Keeling worked feverishly to meet his deadline, the joint article by Revelle and Suess appeared. Though it was little noticed beyond their circle of peers, the paper would one day resound like the tocsin of the environmental movement. After taking into account the pioneering studies of Arrhenius, Callendar, and others, then comparing them with their own data, the authors concluded that "human beings are now carrying out a large scale geophysical experiment of a kind that could not have happened in the past nor be reproduced in the future. Within a few centuries we are returning to the atmosphere and oceans the concentrated organic carbon stored in sedimentary rocks over hundreds of millions of years. This experiment, if adequately documented, may yield a far-reaching insight into the processes determining weather and climate. It therefore becomes of prime importance to attempt to determine the way in which carbon dioxide is partitioned between the atmosphere, the oceans, the biosphere and the lithosphere." The authors' prefatory abstract even contained an oblique reference to Keeling: "An opportunity exists during the International Geophysical Year to obtain much of the necessary information."

Only now did it dawn on Keeling that something much greater was at stake than his obsession with measuring an invisible, odorless, incombustible gas. "In several respects," he wrote in a memoir of the period, "Revelle's interest in carbon dioxide was parallel to mine, but his awareness of human intervention on a global scale and his enthusiasm for the 'far-reaching insight,' which this intervention might promote, were exciting ideas which placed atmospheric carbon dioxide in a new perspective."

Keeling made the final adjustments to the latest version of his manometer, then personally crated and loaded the instrument aboard a ship bound for Antarctica only hours before its departure from San Diego harbor. It was Christmas Eve, and he missed most of the party at Scripps. Worse yet, his fanatic attention to detail proved insufficient once the instrument reached Little America, the base established by Admiral Richard Byrd in 1929. The analyzer's pumps immediately began leaking, and the readings it provided were nothing short of "gibberish."

In early 1958, Keeling's second manometer was bound for the slopes of Mauna Loa, a 13,680-foot volcano in the south central part of the island of Hawaii. He had chosen this location for two reasons: It was the site of a station run by the U.S. Weather Service, and the trade winds that waft across the Hawaiian archipelago carry some of the least contaminated air on Earth. An employee of the Weather Service installed the instrument in March, and it was soon providing Keeling with readings via a needle and graph. The first number that came up was 314, virtually identical to the results of his expeditions into the mountains of California.

His initial triumph was short-lived. Sporadic and extended power outages were interspersed by readings that were all over the place, first up, then down, then up and down again. Equally maddening was the fact that Keeling himself was thousands of miles away, with no chance to lay soothing hands on his "baby."

One problem was resolved when the Weather Service put up the money for new electric generators. With the power failures at an end, a deeply puzzled Keeling watched the concentration of CO_2 rise throughout the winter, until it peaked at 318. Then, with the coming of spring in the Northern Hemisphere, it began to decline, bottoming out at the original measurement of 314. What was one to make of this?

When the time came to plot the first year's data, the reason behind the fluctuating numbers surfaced with a liberating clarity. The rise and fall of the CO_2 curve matched the annual carbon cycle in the biosphere. During the spring and summer, trees and other vegetation take in CO_2 from the air, thus producing the low points on the curve. In autumn, when nature strips trees of their leaves, rendering them dormant, the decaying matter releases its stored carbon and the curve rises once more. What Keeling had inadvertently plotted was nothing less than the rhythmic breathing of the planet.

In retrospect, the results should not have surprised him, for he had long since observed similar fluctuations in readings taken during the same twenty-four-hour period. Each morning, as the Sun rises, photosynthesis results in a gradual drop in the amount of atmospheric CO_2, which bottoms out in the late afternoon. At sunset the process is reversed, as plants release CO_2 throughout the night, though in slightly smaller amounts than they have taken in to support growth. This daily cycle is but a microcosm of the seasonal round.

Early on in his career Keeling had vowed to himself that he would only take a job that would allow him "to study nature itself, not merely ways of exploiting it." He had paid the piper long since, and now his turn to dance had finally come.

13

PENDULUM

~

I think no scurrying will help our cause. Our destined bones
will line
old gravel pits. Mammoth or saber-tooth or man,
there'll be no difference then.
Our interglacial summer passes quickly on.

—Loren Eiseley,
"Why Does the Cold World Haunt Us?"

Kingdom Animalia, phylum Chordata, subphylum Vertebrata, class Mammalia, order Proboscidea, family Elephantidae, genus *Mammuthus*, common name mammoth. Shaggy citizen of the Pleistocene, it was about the size of the modern elephant and roamed every continent except Australia and South America from upwards of 2 million years ago to the last retreat of the glacial ice around 10,000 B.C.

Among the best-known species is the woolly mammoth of Eurasia and North America, which thrived in an arctic climate along the fringes of the vast ice sheets, feeding on coarse tundra vegetation with the aid of complex molars containing as many as twenty-seven traverse ridges. Standing ten feet at the shoulder, it had a compact sloping body that was protected from the extreme cold by a three-and-a-half-inch layer of fat, a rich coat of dense fine hair, and a long outer coat of coarse reddish fiber. While its heat-conserving ears were among the smallest in the elephant world, it boasted a huge pair of tusks that spread outward from the jaw before curling back on themselves, making them useless as weapons but a great aid in breaking through the crusted snow to forage.

Paleolithic man knew the mammoth well. Awed Picassos

of the cave captured its likeness in bold strokes on rock walls, then chased the great beast over cliffs in quest of meat and ivory. Some mammoths blundered into ice crevasses, where their bodies were remarkably well preserved. Instances have been reported of Siberian sled dogs feeding on the flesh cut from mammoth carcasses frozen for 30,000 years, while the plentiful tusks were exported to China beginning in medieval times.

Largest of the large was the imperial mammoth (*M. imperator*), which migrated to North America during the late Pleistocene and thrived in the southern Great Plains. Standing thirteen and a half feet at the shoulder, this earth shaker was faced down by another migratory species armed with only spear and arrow. Indeed, some paleontologists theorize that human predation rather than environmental change was responsible for the mammoth's eventual extinction, though rumors of its existence persisted well into the nineteenth century. President Thomas Jefferson was hopeful that Lewis and Clark might sight and kill such a beast during their epic quest for the Northwest Passage, little knowing that the last of its kind had long since died alone, trumpeting into the void.

~

The mammoth is gone forever, but scientists are less certain about the arctic winter in which the beast flourished. Indeed, just when it appeared that research undertaken by early advocates of the greenhouse effect was on the verge of gaining a wider hearing, the climatic pendulum swung once more: Fire, it seemed, was about to be eclipsed by ice.

The steady rise in temperatures beginning in the 1880s suddenly peaked about 1940, after which the curve began a pronounced descent. This was not to be a temporary phenomenon. Although the level of decline moderated a decade later, the cooling trend lasted for about a third of a century. England's growing season shrank by nine or ten days between 1950 and 1966, while in the northern tier of the Midwestern United States summer frosts occasionally damaged crops. Sea ice returned to Iceland's coasts for the first time in more than forty years, and glaciers in Alaska and Scandinavia slowed their retreat. By the late 1970s, the lion's share of Callendar's estimated one degree Fahrenheit of temperature increase had been eaten away in apparent vindication of his critics.

Nobody at the time could come up with a satisfactory explanation, nor is there a commonly accepted one today, although hypotheses abound. Paradoxical as it may seem, one of the most compelling theories holds that volcanoes were fighting off the heat.

Scientists have long known that a significant cooling of the atmosphere follows eruptions of the magnitude of Krakatau and its ten times more powerful cousin, Mount Tambora, whose monumental explosion in 1815 killed 100,000 at nearby Sugar in the Java Sea. Halfway around the world, frost blighted the life-sustaining potatoes sewn by Irish peasants, while New England farmers endured "a year without a summer," forcing more than 10,000 of them to flee Vermont and Maine for warmer surroundings, a Dust Bowl hegira in reverse.

Borne on thermal updrafts, volcanic gases ride the high stratosphere, where they form bright droplets of sulfuric acid capable of reflecting as much as 2 percent of the Sun's light back into space, cooling Earth by a few degrees. While the mid–twentieth century experienced no eruptions of the magnitude of those that rocked the Java Sea like a cooking pot, scientists at the University of Wisconsin determined that Earth's combined volcanic discharge was about double the average between 1945 and 1975. Hence the warming effect attributed to rising carbon dioxide levels may have been temporarily overridden by this natural phenomenon, making it appear that Arrhenius and his disciples had simply gotten it wrong.

Another hypothesis attributes such global temperature fluctuations to the Sun, which the Earth orbits from a mean distance of 92,960,000 miles. Some 865,000 miles in diameter, the Sun has a volume 1.3 million times that of Earth, and a mass 700 times greater than all of the other bodies in the solar system combined. At its nuclear core, under the enormous pressure of a billion atmospheres, it generates temperatures reaching 10 to 20 million degrees Celsius, which taper off to a mere 6,000°C at its surface, otherwise known as the photosphere. Yet for all this, we now know the Sun to be an inconstant star.

Ever since the nineteenth century, mostly unsung solar astronomers have maintained lonely vigils on desolate mountain peaks collecting data by the hour, the day, the week, the month, the year. But always there were the distorting effects of the atmosphere—dust, clouds, air currents, pollution, and a host of other variables beyond the scientist's control. Not until Valentine's Day, 1980, when NASA boosted a satellite nicknamed Solar Max into orbit 400 miles above Earth, did conclusive data begin coming in. With its unobstructed view of the star, Max's highly sensitive light meter recorded day-to-day variations as large as 0.25 percent. As the data accumulated, scientists also detected a general trend. Between 1980 and 1985, the amount of sunlight reaching Earth's atmosphere decreased by an average of .019 percent per year. Further calculations indicated that even small fluctuations may make a big difference insofar as human beings and other species are concerned.

The wheels were turning as climatologists gleaned the historical record for possible correlations. Among the most promising was the Little Climatic Optimum, when the Anasazi flourished in the desert Southwest while Erik the Red and his Viking band colonized the western fringes of Greenland. Reaching back even farther, they theorized that the world's first major civilizations, including Sumeria, Minoan Crete, India, and Egypt, matured under the benevolence of a kindly Sun around 3000 B.C., when global temperatures rose one or two degrees Celsius.

Conversely, what the Sun gives it can just as easily take away. There is evidence that it may have dimmed several times in the past 10,000 years, most conspicuously during the Little Ice Age that witnessed the demise of the Anasazi and the Norsemen. Twice in the early fourteenth century, the Baltic Sea froze fast and glaciers came out of hibernation, grinding their way south to a point last reached 15,000 years earlier. Londoners roasted oxen on the Thames while the Flemish artist Pieter Brueghel, his fingers stiffened by the cold, kept up his spirits by painting peasants cavorting in the snow. At Plymouth Colony, a Pilgrim wrote of the "cruell and fierce [winter] stormes" that racked the settlement; 200 years later iceboats plied the Hudson River almost as far south as New York City.

According to Solar Max, the Sun reversed itself and began

to brighten again in 1986, though the precise reason is a mystery. Yet if such cycles are even partially responsible for the making and breaking of civilizations, humanity is likely to be in for some interesting times. Should the Sun's energy output hold steady or slightly decline as carbon dioxide levels continue to rise, we may have dodged a potentially lethal bullet. It is even possible that we will have inadvertently served our cause by adding to the insulating blanket surrounding Earth, thus staving off the natural return of the glaciers. If, on the other hand, the Sun grows warmer, the piggyback effect of rising CO_2 levels could result in a devastating scenario. As Jonathan Weiner, the noted science writer, has observed, "In the next hundred years, it would take a smaller change than ever to destroy civilization, if the change happens to be in the wrong direction."

Like the historian Arnold Toynbee, who read the advance and decline of civilizations in great cycles composed of lesser cycles, climatologists are not without their grand mechanisms. The most audacious of these was set forth by the Serbian mathematician Milutan Milankovitch in the 1920s. Ever since Isaac Newton's revolutionary work on orbital mechanics, scientists have known that Earth leads a merry dance around the Sun. During its travels, it tilts at a slight angle and spins on its axis like a giant top overloaded on one side, so that the gravitational pull of the Sun and Moon acts on it asymmetrically. This causes Earth's axis to wobble and describe a cone every 26,000 years, a movement familiar to astronomers as the "precession of the equinoxes." It simply means that over millennia, the planet is closest to the Sun at different times of the year. Other bodies also get into the act, especially Jupiter, the largest and most massive of the planets. Its pull causes Earth to stagger, or "perturb," as it orbits, factors that gave Milankovitch an idea.

Perhaps these cycles interact in a complex pattern to vary the amount of solar radiation the Earth receives. It stands to reason that during periods of lesser radiation summers will be cooler, causing northern snows to linger, triggering another ice age and the snail-like advance of glaciers. Conversely, greater solar radiation would eventually release the planet from the north wind's deadly grip.

Based on thirty years of calculations, Milankovitch concluded that the drop in the amount of sunlight reaching Earth at the extreme end of both the precession and tilt cycles could cool a hemisphere just enough to marshal the ice. He further determined that the largest ice sheets recur in 100,000-year cycles—"Milankovitch cycles"—punctuated by alternating warm and cold periods, each lasting about 10,000 years.

Milankovitch's orbital calculations have since gained the support of other scientists. Moreover, data gathered from such far-flung sites as the polar caps, Macedonian peat bogs, and coral reefs indicate that there was a sudden shift back to warm weather about 10,000 years ago. Notwithstanding other periods of warming, ours is the most benign cycle in 100,000 years, yet Earth's orbital dance may be signaling its approaching end. The last glaciers pulled back from much of Scandinavia about 8,000 years ago, while the peak of the last ice age was reached almost exactly 10,000 years before that. Whatever happens in the coming centuries, it appears that this wobbling, tilting, staggering planet will eventually transport us back down an icy road, though precisely when is uncertain. Nor did Milankovitch live long enough to factor in another element that may yet compose a variation on his grand symphony—human intervention.

If volcanic eruptions, the waning of the Sun, or Milankovitch cycles were found wanting as explanations for nearly thirty years of cooling, there were plenty of other theories available. While scientists were scratching their heads as temperatures continued to drop in the mid-1960s, oceanographers David B. Ericson and Goesta Wollin wryly noted, "It has been estimated that a new theory to explain continental glaciation has been published for every year that has passed since the first recognition of the evidence for past glaciation." The puckish British climatologist C. E. P. Brooks simply quoted Kipling:

> *There are nine and sixty ways of constructing*
> *tribal lays,*
> *And every single one of them is right.*

Added Brooks, "There are at least nine and sixty ways of constructing a theory of climate change." However, not even in jest did Brooks claim that every one of them is right.

The press had a field day echoing the dire prognostications of scientists. "There are ominous signs," an article in the April 28, 1975, issue of *Newsweek* began, "that the earth's weather patterns have begun to change dramatically, and that these changes may portend a drastic decline in food production—with serious political implications for just about every nation on earth. The evidence in support of these predictions," it continued, "has now begun to accumulate so massively that meteorologists are hard-pressed to keep up with it." Temperatures during the great ice ages were only about seven degrees Fahrenheit lower than during Earth's warmest eras, and the recent decline had taken the planet about one-sixth of the way toward the lost world of the mammoth.

The chances that politicians would take any action seemed remote, even though scientists had come up with a number of strategies to revitalize the cooling world. One proposal called for spreading soot over the arctic ice cap to decrease its albedo, or reflectivity; another favored diverting arctic rivers in an attempt to raise the temperature of the ocean currents. It was admitted that such tinkering might create problems far greater than those they hoped to resolve, yet few government leaders were so much as willing to stockpile food or introduce the variables of climatic uncertainty into projections of future food supplies. "The longer the planners delay," author Peter Gwynne cautioned, "the more difficult will they find it to cope with climatic change once the results become grim reality."

Predictions of a catastrophe in waiting coincided with the issuance by the Hugh More fund of a pamphlet in whose pages the term "population bomb" first appeared. In 1968 Dr. Paul R. Ehrlich of Stanford University's Department of Biological Sciences borrowed the compelling phrase, using it as the title of his impassioned lament on a species replicating itself to disaster. "Man can undo himself with no other force than his own brutality," Ehrlich began. "The roots of the new brutality . . . are a lack of population control." Unlike the prophesied return of the ice, the catastrophe of which the author wrote was at hand: "The battle to feed all of humanity is over. In the 1970's the world will undergo famines—hundreds of millions of people are going to starve to death in spite of any crash programs embarked upon now." And so they did, if not in the staggering numbers predicted by Ehrlich.

On another front, the alternately acerbic and eloquent environmentalist Edward Abbey was dreaming up his fictional alter ego cum hero, Vietnam veteran George Washington Hayduke, who has had it with strip miners, clear-cutters, and the highway, dam, and bridge builders. Hayduke forms the seriocomical Monkey Wrench Gang (the eventual title of Abbey's novel), a bizarre quartet of ecoraiders who vow to halt the construction of a road across the Arizona desert by acts of creative sabotage.

While Doc Sarvis, a wild conservative and libertarian, stands watch, his three comrades busy themselves "cutting up the wiring, fuel lines, control link rods and hydraulic hoses of the machine, a beautiful new 27-ton tandem-drummed yellow Hyster C-450A, Caterpillar 330 HP diesel engine, sheepsfoot rollers, manufacturer's suggested retail price of $29,500 FOB Saginaw, Michigan. One of the best. A dreamboat." Then Hayduke delivers the coup de grâce by unscrewing the oil-filler cap and pouring sand in, while the outcast Mormon and polygamist Seldom Seen Smith empties four quart bottles of sweet Karo syrup into the fuel tank. "The engine would seize up like a block of iron, when they got it running. If they could get it running." Bonnie Abbzug, an exile from the Bronx, wants to slash the seats for good measure, but Doc angrily draws the line. "That's vandalism. I'm against vandalism. Slashing seats is petty-bourgeois."

Life was soon imitating art, indeed doing fiction one better. A frustrated Dave Foreman, the chief Washington lobbyist for the Wilderness Society, departed the capital and together with a few friends formed Earth First!—a radical environmental group. Its motto is "No compromise in defense of Mother Earth"; its symbol the monkey wrench raised in defiance. The group's journal was soon providing tips on sabotage—or what it called "ecodefense"—that included everything from disabling heavy machinery with a mixture of one part silicon carbide to four parts motor oil to driving iron spikes into trees about to be felled by chain saw–wielding loggers. The now familiar arrests for acts of civil disobedience quickly followed.

Meanwhile, the National Geographic Society was transforming French oceanographer Jacques Cousteau into an international icon. Plying the waters of the world on his beloved *Calypso*, the inventor of the Aqua-Lung found evidence of man-made pollution everywhere he sailed, no matter how far from port, no matter how deep, thus belying Grace Hubble's notion that human beings could do nothing to harm the sea. Into the gaping breach sped Greenpeace, its tiny rubber craft zipping across the bows of supertankers and nuclear vessels with suicidal abandon.

Like a blow on a bruise, the little-known second law of thermodynamics was suddenly thrust into public consciousness by the technological gadfly Jeremy Rifkin, author of the best-selling jeremiad *Entropy*. After flaying Isaac Newton and his fellow natural philosophers for being the architects of massive energy consumption in the industrial age (of which they were ignorant), Rifkin sounded the alarm: "The second law, the Entropy Law, states that matter and energy can only be changed in one direction, that is, from usable to unusable, or from available to unavailable, or from ordered to disordered." Burn a lump of coal, for example, and its energy is transformed into sulfur dioxide and other gases that rise into the atmosphere. While no energy is lost because the first law of thermodynamics states that all matter and energy in the universe are constant—that they can be neither created nor destroyed—neither can the lump of coal ever be burned again; its dissipated energy has become useless in terms of accomplishing human tasks. And because the amount of available energy is finite, the human clock, the planetary clock, the clock governing the universe itself are all ticking off the millennia to that final moment when the last feeble star flutters and blacks out.

~

Not once in the intervening twenty years had Charles Keeling deserted his watch. Having found a permanent home at Scripps, he continued to test the atmosphere for carbon dioxide with the same fanatical precision of his younger days. At first, the world's CO_2 rose about one part per million per year. It accelerated to one and a half parts per million, then two, then three, never once leveling off, let alone dropping, although the rate of rise slowed briefly in 1973 due to the oil embargo. Half jokingly, the chemist later confessed that he felt personally responsible for what was happening.

The upshot was the "Keeling curve," arguably the most famous graph in all of Earth science. With the year 1958 and a CO_2 level of 315 ppm as its base, the curve plots a global rise of about 11 percent through 1997, when the CO_2 level stood at some 365 ppm. Moreover, the concentration is projected to increase, like compound interest, at a rate of 0.5 percent a year, perhaps more, depending on what China and India do with their massive coal reserves. Herein lay the proof that believers in human-induced global warming had been searching for, the underlying engine of the greenhouse effect.

Keeling's findings have been bolstered by the analysis of ice cores, which yielded a concentration of about 280 ppm in the late nineteenth century, a figure consistent with the climatological data compiled by George S. Callendar. And like Cal-

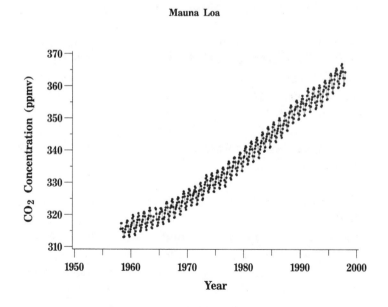

The Keeling curve showing CO_2 readings
from the Mauna Loa Observatory
(Dave Keeling and Tim Whorf, Scripps Institution of Oceanography)

lendar, Keeling played the part of the historian by tracing the consumption of fossil fuels back to 1850, when CO_2 levels recorded a doubling time of fifteen to twenty years, with but one modest exception. A slowing down occurred during the decade of the 1930s because of the worldwide depression. Whether coincidental or not, this tapering off preceded the thirty-year decline in temperatures that others attributed to volcanic activity, Milankovitch cycles, and other natural phenomena such as interstellar dust, a shift in Earth's magnetic field, the surging of the antarctic ice sheet, and the deep circulation of the oceans.

Keeling's work has also gone a considerable way in helping to resolve the issue raised by Hans Suess and Roger Revelle in their 1957 article on CO_2 and the oceans of the world. That these bodies of water act as sinks for CO_2 is beyond dispute, but that they are slow sinks now seems disturbingly clear. Of the billions of metric tons of CO_2 emitted into the atmosphere during the industrial revolution, a large portion remains there—molecules from Coalbrookdale's tall chimneys presided over by the fifth generation of Darbys; from Henry Bessemer's exploding converters and Andrew Carnegie's hell-fired steel mills; from John D. Rockefeller's flaring oil fields; from the exhaust of Henry Ford's first Tin Lizzie, and his last, and from all that followed. The buildup had taken place over generations, and it suddenly became obvious that it would take generations more to stabilize CO_2 levels, let alone bring them back down, and only then if the citizens and politicians of the world were convinced that such drastic action was imperative.

~

Still another storm was brewing on the environmental front, kicked off by a paper in the June 1974 issue of *Nature* under the joint authorship of F. Sherwood Rowland, a professor of chemistry at the University of California, Irvine, and Mario J. Molina, a postdoctoral fellow. They began by noting that chlorofluorocarbons were being added to the environment in steadily increasing amounts and could be expected to reach concentrations ten to thirty times present levels. Equally worrisome was the fact that these compounds, which were the brainchild of Thomas Midgley Jr., are chemically inert and may remain in the atmosphere anywhere from sixty to one hundred years. Then the bombshell: "Photodissociation of the chlorofluoromethanes in the stratosphere produces significant amounts of chlorine atoms, and leads to the destruction of the atmospheric ozone."

The most chemically active form of oxygen, ozone is an unstable, bluish, poisonous gas with a distinct acrid odor. It is used commercially as a disinfectant and decontaminant for air and water, and as a bleaching agent for waxes, oils, and other organic compounds. Ozone is also formed naturally thirty miles above Earth in the cold and barren stratosphere by the action of solar ultraviolet light on oxygen. Although it is present in the "ozone layer" to the extent of only about ten parts per million, it is critical to all life below, for ozone blocks most ultraviolet rays harmful to foliage, insects, birds, and the cancer-prone skin of humans. Scientists believe that there was no life on the continents before there was ozone to protect DNA molecules, though the ocean depths may have been safe for the early development of living things.

Rowland and Molina described how, once the CFCs reach the stratosphere, they are broken apart by ultraviolet radiation. The result is the release of chlorine, a poisonous gas and asphyxiant, like carbon monoxide and carbon dioxide. Although it was discovered in 1774, chlorine did not become known to the public at large until the spring of 1915, when the German army, desperate to break the slaughterous stalemate on the Western Front, used it in the battle of Vimy Ridge. Casualties soared, stirring outrage among the civilian population. Months later British soldiers retaliated in kind by launching chlorine canisters into the German lines, also to no avail.

After its release, the chlorine attacks and destroys the ozone in a chain reaction known to chemists as a catalytic cycle. It takes only a matter of seconds for the process to complete itself, but it does not end there. The chlorine atom immediately attacks another ozone molecule, destroying it in turn. It is estimated that a single chlorine atom can repeat the catalytic cycle some 1,000 to 10,000 times before it finally combines more permanently with hydrogen, causing it to stabilize.

Midgley's CFCs, long viewed as miraculous for their superb qualities as refrigerants and, more recently, as aerosol propellants, were destroying a life-giving band of molecules so insubstantial as to render gossamer leaden by comparison. Hair spray, deodorants, furniture polish, abandoned refrigerators, each the banal product of industrial civilization, suddenly threatened their creators with deadly carcinomas and blinding cataracts, a modern cross between the biblical leper and Sophocles' Teiresias.

The oncoming debate would be intensified by the decision of manufacturers on both sides of the Atlantic to build a sleek new race of aircraft dubbed the Supersonic Transport, or SST. Soaring above the troposphere at twice the altitude of conventional passenger jets, the SST would spew nitrogen oxides from its engines that would linger in the stratosphere indefinitely. If significant numbers of the aircraft entered regular service, the ozone could be reduced by an additional 10 to 20 percent. Even more troubling to some scientists was the prospect of the recently designed space shuttle tearing through the stratosphere, its gargantuan engines at full throttle.

Rowland and Molina's article had been in print only eight months when another one on CFCs appeared in the October 1975 issue of the journal *Science*. Its author was a young atmospheric chemist named Veerabhadran Ramanathan, who had recently completed his doctorate at the State University of New York at Stony Brook. Ramanathan's interest in CFCs stemmed from his dissertation research. He was able to demonstrate that CFCs trap the Sun's energy in the same manner as do the other greenhouse gases, such as water vapor, carbon dioxide, and methane. In fact, molecule for molecule, CFCs are much more efficient at trapping infrared radiation than the other greenhouse gases. "This enhancement," Ramanathan warned, "may lead to an appreciable increase in the global surface temperature if the atmospheric concentrations of these compounds reach the values of the order of 2 parts per billion." Ozone depletion and a potential increase in the greenhouse effect. It was this dual portent of creeping disaster that was about to touch off what environmental historians Lydia Dotto and Harold Schiff christened "the Ozone War."

~

As if atmospheric scientists did not have enough on their plates already, they were suddenly faced by another conundrum in the mid-1970s. After some thirty years of cooling, the climatic pendulum seemed momentarily suspended, as though in equipoise; then it abruptly reversed itself, causing global temperatures to rise once again.

Writing with the same conviction as those who had read Earth's future in ice, proponents of the global warming theory were led by Wallace S. Broecker, an ocean chemist on the staff of Columbia University's Lamont-Doherty Geological Observatory. In 1975, with the planet only a year away from its next warming cycle, Broecker wrote in *Science* that "a strong case can be made that the present cooling trend will, within a decade or so, give way to a pronounced warming induced by carbon dioxide. . . . Once this happens, the exponential rise in the atmospheric carbon dioxide content will tend to become a significant factor and by early in the next century will have driven the mean planetary temperature beyond the limits experienced during the last 1000 years." Broecker even went so far as to charge those who discount the warming effects of industrial CO_2 with "complacency."

Ironically, the chemist's data on warming derived from ice core samples collected in Greenland. Never, so far as this frozen record was concerned, had Earth remained cool when the level of CO_2 in the atmosphere was high, and Keeling's curve placed it at a robust 330 parts per million at the time Broecker was writing. Conversely, there was no record of an ice age except when CO_2 levels had dipped below some 200 ppm, and the possibility of that occurring anytime soon seemed far-fetched.

Strengthening the argument was a paper that appeared in *Nature* within a week of Broecker's. Its authors were two New Zealand scientists, M. J. Salinger and J. M. Gunn. They reported that while the Northern Hemisphere had been cooling for decades, southern latitudes had been warming over the past thirty years, though they ventured no opinion as to whether this trend would push northward across the equator. It was almost as if Dante had somehow anticipated it all back in the thirteenth century: "Into eternal darkness, into fire and into ice."

14

A DEATH IN THE AMAZON

~

Hurt not the earth, neither the sea, nor the trees.

—Revelation

A Culture is no better than its woods.

—W. H. Auden, "Bucolics"

A very long time ago, a mountainous archipelago and a continental shield were joined at the hip by nature, then bonded by an ice cap thousands of feet thick to form Antarctica, Earth's fifth-largest continent and its highest. The steeply sloping ice is forever fracturing and breaking away at the seaward margins, floating off like miniature mountains and leaving behind towering blue-and-white cliffs hundreds of feet high. During winter, as sea ice forms at its periphery, the continent doubles in size, sprouting enormous frozen tongues that project into the stormiest waters on the globe. Locked within its vast reaches is 70 percent of the world's freshwater, the purest anywhere in nature, yet the continent is so dry that geologists classify the 5 percent of its surface free of ice as desert.

It is a continent with no native human population, nor any large land animals, a place whose interior is the coldest on Earth. On August 24, 1960, at the Soviet Union's Vostok Station, the temperature dropped to −126.9ºF (−88.3ºC), the lowest ever recorded—anywhere. Added to this are fierce interior winds that race along the ice-drowned mountains at speeds as great as 200 miles per hour, creating windchill conditions for which no realistic human scale exists.

It was into this white void that sixteen expeditions representing nine countries came in the 1890s, financed by private individuals and scientific societies. Two decades later Ernest Henry Shackleton sledged to within ninety-seven miles of the South Pole, but it was Roald Amundsen who, on December 14, 1911, carried the day by planting the flag of his native Norway atop the Holy Grail of the underworld. His dash by dog team and skis bested the heartbroken Robert Falcon Scott, who arrived thirty-five days too late. Having returned to the safety of his ship on the Bay of Whales, Amundsen had no way of knowing that his rival's starving and frostbitten party had perished in a blizzard only a few miles short of their base on the Ross Sea.

Hundreds of miles to the north of Scott Base lies Halley Bay, outpost of the British Antarctic Survey and for decades the periodic residence of Dr. Joseph Farman, Charles Keeling's doppelgänger. Like the steadfast chronicler of atmospheric carbon dioxide, Farman had been an eager young Ph.D. when the International Geophysical Year was launched in 1957. He first arrived on Antarctica's frozen shore in September, one of the few periods when the austral Sun neither rises nor sets, but simply moves around the horizon in a circle, bathing the coruscated landscape in perpetual light. The end of this reprieve is signaled by the departure of the supply ships, after which the Sun sets and does not rise again for months, plunging the continent into perpetual darkness.

Farman, a man of medium build given to rapid speech and nervous mannerism, was assigned the unglamorous task of taking measurements of the frigid atmosphere. He worked with an instrument called a spectrophotometer, which had undergone little change since its invention in 1926. One of the things he was required to track was the level of ozone, and he dutifully took his first readings not long after he hit "the ice," framing patterns of light penetrating the molecules of Halley Bay.

The British Antarctic Survey was little known within the scientific community, as was its obscure outpost. When Farman returned to England and Cambridge as the IGY was winding down, the prospects for additional funding were uncertain at best. Yet there was something about the ice that called him back. He had no illusions about glory of the kind bestowed on Newton and Darwin; rather, like Keeling, he simply wanted to provide the rest of science with a reliable baseline of atmospheric data stretching over decades.

With a tenacity put to use, Farman ran the funding gauntlet year after year, somehow managing to squeeze enough money out of his employer, the Natural Environmental Research Council, to head south each September. In 1981, after twenty-five years of continuously monitoring the polar skies, he was still earning the equivalent of less than $20,000 a year, but he had additional compensation in the form of an unbroken line of data to show for his sacrifice.

Farman kept close track of the battle launched by Rowland and Molina's 1974 article in which they theorized that CFCs were destroying the atmospheric ozone. Lacking proof in the form of instrument readings, the authors were subjected to attacks both within and without the scientific community. Particularly vicious were the assaults by the chlorofluorocarbon industry, whose sales from their manufacture had climbed into the billions. In a scenario that presaged the bitter and prolonged tobacco war, the researchers' conclusions were branded as "preposterous" by the spokespeople of industry, who responded to concerns about skin cancer and crop damage by trotting out an impressive array of corporate chemists, lobbyists, and public relations experts. Their message was unmistakable: Woe to any scientist who dared come to the defense of the heretical pair.

For twenty-seven years running Joe Farman had unloaded his equipment beneath a wan antarctic Sun, but in 1984 he remained in Cambridge, leaving the fieldwork to a younger generation of enthusiasts. He had delayed publishing the results of his fieldwork long enough, three years to be exact, hoping against hope that some other scientist (the Japanese and Americans were also on the ice) would publish the secret that lay hidden in the reams of data spread across his battered desk.

Farman had been highly skeptical when the September 1981 readings from his Halley Bay instruments showed that the amount of ozone had declined during what passed for springtime in the Southern Hemisphere. He suspected that his aged spectrophotometers were out of calibration and asked his assistants to check them over. After a dip of some sixty Dobson units (20 percent below normal), the readings returned to their baseline level. Curiously enough, the same thing happened in 1982 and again in 1983, except that the decline grew progressively worse. By 1984 the springtime ozone levels in the stratosphere were a staggering 40 percent below average. Metaphorically speaking, there was a hole in the sky—an "ozone hole."

The lonely watchman of the lonely continent had stumbled onto something far greater than he had bargained for, as had Keeling when he first measured the respiration of the biosphere. Moreover, a careful analysis of the data indicated that the first serious glitch in ozone levels had actually occurred in 1977, only three years after Rowland and Molina published their findings in *Nature*. With so much at stake, Farman could no longer maintain his silence in good conscience. He assembled the survey's advisory group, and the decision was unanimous: Publish as soon as possible, and let the chips fall where they may. Farman recollected going home one evening and telling his wife, "The work is going well, but it looks like the end of the world."

Bearing the recondite title "Large Losses of Total Ozone in Antarctica Reveal Seasonal CIO_x/NO_x Interaction," the finished manuscript was submitted to *Nature*. By now the ozone hole had gotten so big that it reached nearly to Tierra del Fuego at the tip of South America, where another British monitoring station detected it, much to Farman's relief. Then came word that the prestigious National Academy of Sciences had issued its fourth and most optimistic report on the ozone layer. New calculations indicated that it would decline by only a few percent during the coming century and might even undergo a slight rise. Farman began having visions in which he was subjected to the guffaws of his peers for publishing "bad science."

Not only did Farman believe that the seasonal depletion of ozone had reached critical levels, he had a theory about why it was happening over the South Pole. Very low temperatures coupled with the seasonal peak in ultraviolet radiation render the antarctic stratosphere uniquely sensitive to the growth of inorganic chlorine. The good news was that comparable effects should not be expected in the Northern Hemisphere, where the polar stratosphere is neither as cold nor as stable as its southern counterpart.

Farman's article was published on May 16, 1985, but it was the National Academy of Sciences report that grabbed the headlines. *Science Digest*, one of several publications that covered the story, titled its offering "Ozone: The Crisis That Wasn't." When Sherwood Rowland, who felt himself vindicated by Farman's work, spoke on the University of Maryland campus in November, not one of the eastern papers acknowledged the press release by sending a reporter.

By now Farman was willing to accept any help he could get, including intervention from on high. While his ground-based spectrophotometer was pointed upward toward the ozone layer, a NASA satellite named *Nimbus 7* was peering down from the firmament and measuring the ultraviolet light being reflected back into space, a process that depends on the amount of ozone present. The silver-winged seraph, which had been launched in late 1978, was even equipped with a sophisticated instrument known to meteorologists as TOMS, for Total Ozone Mapping Spectrometer. Orbiting Earth via the poles once every hour and a half, *Nimbus 7* radioed the TOMS data on ozone back to the planet, where it was instantaneously classified as gibberish by a computer programmed to disregard numbers that fell outside certain parameters.

After Farman published, NASA scientists became suspicious and decided to go back into *Nimbus 7*'s data archive to see if they had missed something. Sure enough, the computer had been told not to factor in numbers lower than 180 Dobson units because nobody had dreamed that an ozone hole might exist. The computer was reprogrammed, and the result was chilling. *Nimbus 7* had actually detected the ozone hole well before Farman and his colleagues discovered it with the aid of their primitive instruments. Thomas Midgley's brainchildren had been up to serious mischief all along.

~

On an April day in 1974, the leaden skies over Pitlochry, Scotland, suddenly opened up, drenching the town and the surrounding countryside. There was nothing unusual in this, for rain was as common as the heather whose emerging white and rose and yellow flowers signaled the return of spring to the ancient hills and glens. However, it soon became obvious that this gray downpour was like no other. A sample was found to be 1,500 times more acid than normal rainwater, as if vinegar were falling from the sky.

Had Robert Angus Smith still been alive, this finding would not have surprised him, except perhaps in its magnitude. A chemist by training, Smith served as Britain's first alkali inspector in accordance with an act passed by Parliament in 1863. A decade before that, he had discovered a link between the soot-laden skies over industrial Manchester and the high level of acidity in the region's precipitation. In 1872 Smith published his now classic study, *Air and Rain: The Beginnings of a Chemical Climatology*. Pitlochry, like Manchester and Glasgow and London before them, had been subjected to an involuntary immersion in what Smith called "acid rain."

Normal rain, with a pH of about 5.6, is of itself slightly acidic, a condition that results when carbon dioxide from the atmosphere reacts with moisture in the air to form carbonic acid. The formation of acid rain, which has a pH of 5.5 or lower, begins with atmospheric emissions of sulfur dioxide and nitrogen oxide from automobiles, industrial operations such as smelting, and power-generating plants that burn fossil fuels. These gases combine with water vapor in the clouds to form sulfuric and nitric acids, powerful corrosives that can pit the metal girders of bridges, efface the architectural details of classical buildings, and dissolve outdoor sculpture that has withstood the centuries.

It was hardly by coincidence that industrial Manchester's middle class chose not to build their homes on the east and northeast sides of the city, where for ten or eleven months of the year the wind drove the factory smoke across the rooftops and into the cramped and dreary quarters of the workers. In March of 1872, the year Smith published *Air and Rain*, the *Illustrated London News* attributed the widespread death of vegetation around Liverpool and Newcastle to acid pollution. The practice of resorting to ever taller smokestacks had done no good. "The erection of very high chimney shafts was useless, and only disseminated the evil over a larger area."

Pollutants that combine with water vapor to form acid rain are typically borne hundreds and sometimes thousands of miles on the prevailing winds. Little of this was known in the 1920s and 1930s, when the lakes of Scandinavia underwent a profound change. It started with the disappearance of salmon from southern Norway after World War I. Inland stocks of brown trout also declined so precipitously that 2,000 lakes would be stripped of these and other species in the coming decades.

Neighboring Sweden fared even worse. The crayfish almost vanished, forcing the crustacean-loving Swedes to fall back on imports from Turkey. Other species followed until, by the 1980s, 4,000 lakes were effectively moribund and 5,000 more had sustained heavy fish losses.

By 1959 the connection had been made. Acidification was declared the culprit. Its victims weren't limited to fish populations. Plankton, insects, snails, freshwater shrimp, and other species were fast disappearing as the pH of lakes continued to rise, sealing their fate. The lethal cocktail of toxic chemicals, which included aluminum and cadmium, was also taking its toll on birds such as grebes and diving ducks. As if this were not bad enough, the chemistry of rainwater had been brought to bear on the puzzling question of why tens of thousands of acres of once healthy forests had begun dying in parts of Europe and North America.

The credit for tracing the source of acid rain to Europe's heavily industrialized areas belongs to Svante Oden, a Swedish soil chemist. Envisioning the scourge as a new kind of international "chemical warfare," Oden launched a campaign designed to educate his colleagues to the widening danger. Of major concern was the "tall stacks paradox," dating from the nineteenth century. It involves the substitution of a highly visible but local air problem—namely toxic emissions—for an invisible long-distance transfer of pollutants. Unwitting Victorians believed that the taller the chimney the cleaner the air. But what goes up—and this includes sulfur and nitrogen oxides—must come down eventually, no matter how far afield, a fact of which Oden was keenly aware. He knew whence the components of Scandinavia's acid rainfall had come and rightly singled out Great Britain and the two Germanies.

Ironically, the problem was compounded by a resurgence in tall chimney construction brought on by the passage of pollution control initiatives such as the U.S. Clean Air Act of

1970. For the first time in more than a century, industrial plants and power stations were increasing the height of their smokestacks. By 1981 there were 179 new stacks at least 500 feet tall, 20 of which reached 1,000 feet, more than double the size of the twin giants that had loomed over Glasgow in the 1850s. Another 350 were on the drawing boards for completion no later than 1995.

As Oden struggled against the inertia of "out of sight out of mind," pH levels plummeted. In Kane, Pennsylvania, a rainstorm produced a pH of 2.7, 800 times more acidic than normal and not far from that of lemon juice. This reading was soon eclipsed by one in Wheeling, West Virginia, where a 1979 drenching yielded a record pH of 1.5, the equivalent of stomach acid.

The damage sustained by Sweden and Norway was being duplicated in the United States, where coal-fired power plants in the Ohio and Mississippi Valleys pump tens of thousands of tons of sulfates into the upper atmosphere each year. Riding the winds, the emissions stream into the Northeast to combine with moisture in the air and fall as acid rain. Nowhere are the destructive effects more evident than in the rugged Adirondacks of New York, a circular mountain mass covering 6 million acres of public and private lands that could accommodate Yellowstone, Glacier, Yosemite, and Grand Canyon National Parks, with considerable room to spare.

Despite its wilderness location, the average acidity of precipitation falling on the Adirondacks increased fortyfold between 1930 and 1980. Its thousands of small mountain lakes are especially vulnerable to acidity, for not only do they receive more moisture and hence more acid than most other lakes, they are also poorly "buffered." The soils of the region are thin, and the granitic rocks lack such alkaline substances as calcium carbonate and magnesium carbonate to help neutralize acidity. By the mid-1970s, almost half the lakes registered a pH below 5.0, while fully 90 percent of them had lost their fish stocks. Among them was a once pristine Jockeybush Lake, described by one disconsolate nature writer as the equivalent of a "junked car battery."

As documentation of the decline in aquatic life mounted, scientists became aware of a parallel phenomenon. Spruce and other conifers were dying at high elevations along the Atlantic

Seaboard. Core samples taken of other species revealed that beginning about 1960, the growth rate of a wide variety of trees in the eastern United States had suffered a sharp and continuous decline, yet there had been no prolonged drought or outbreaks of disease to account for this.

Simultaneously, reports began coming in from Germany that the great trees of the Black Forest were dying, also for no apparent reason. At first, the damage was attributed to a baffling and virulent form of "tree cancer," or what German scientists called *Waldsterben*, literally "death of the trees." The first signal that something is amiss comes when the dark green branches of conifers such as the spruce begin to droop, like Spanish moss. Months, sometimes years, later, they become tinged with yellow and then brown. A carpet of discolored needles soon forms beneath the weakened tree, and new needles eventually stop growing. Balding begins at the top and works its way down a warping trunk and branches no longer capable of fending off drought, insects, parasites, and frost, which eventually finish it off. In the end, it stands like a denuded sentinel pointing a gnarled and accusatory finger at the sky.

By 1982 an estimated 5 to 7 percent of the trees in the Black Forest either were damaged or had died; by 1985 that figure had risen to an alarming 50 percent. Although it first seemed that the deciduous species would be spared, they too began to expire. Half the beech and over 40 percent of the oaks were affected, and elms that had a normal life span of 130 years were dying in 60. Once the word was out, foresters in other parts of Europe noted a similar pattern. Scandinavia, France, Britain, Italy, Switzerland, East Germany, Czechoslovakia, Hungary, and the Soviet Union all reported on the ravages of *Waldsterben*.

To the surprise of many, scientists could come up with no hard evidence of a cancerous pandemic. And other likely causes—insects, drought, unusual fluctuations in weather patterns—were eliminated one by one, until researchers were left with a lone common denominator. Everything pointed toward a stygian brew of toxic ingredients, including sulfur dioxide, hydrogen peroxide, nitric acid, and industrially produced ozone. Moreover, an ominous compounding effect was discovered at work. The fog enshrouding many of the world's northern forests is commonly made up of tiny droplets ten times more acidic than the water in acid rain, helping to explain why trees at higher elevations are particularly susceptible to damage.

Scientists also learned that the onset of winter brings no end to the assault, only postponement. Locked within the drifting snows are several months' worth of acidity, all of which is suddenly released during the spring freshet. This outflow merges into streams and lakes without first being partially neutralized by soaking into the ground. And it comes just as fish and reptiles are laying their eggs. Soon the rain and fog return, decimating ever more trees, the very lifeblood of the carbon cycle and the planet's first line of defense against a greenhouse effect that many had begun to fear was mushrooming out of control.

~

In the late spring of 1842, a visiting Charles Dickens, accompanied by his wife, Catherine, and secretary, crossed the state of Ohio by stagecoach on the National Road. From the vehicle's open window, Dickens bore witness to a different and even more wasteful kind of deforestation. The already famous author called this chapter of his *American Notes* "Rough Travelling," as straightforward and literal a title as he could think of. Much of the way was over "corduroy road," which was made by dumping tree trunks into a marsh and leaving them to settle. Dickens noted, "The very slightest of the jolts with which the ponderous carriage fell from log to log was enough, it seemed, to have dislocated all the bones in the human body." He could only liken it to an imaginary attempt to reach the top of St. Paul's Cathedral by omnibus.

At least some comfort could be taken from the fact that the driver was in no danger of falling asleep at the reins, for every now and then a wheel would strike an unseen stump, threatening to dislodge him from his perch. Contractors had been required to level only trees twelve inches in diameter and smaller; eighteen-inch trees could have stumps nine inches high, and those above eighteen inches could have fifteen-inch stumps, rendering the measuring rod as crucial as the ax.

To Dickens, the stumps were "a curious feature in American travelling, astonishing in their number and reality." As twilight descended and the shadows lengthened, the novelist's imagination came alive. "Now there is a Grecian urn erected in the centre of a lonely field; now there is a woman weeping at a tomb; now a crouching negro; an armed man; a hunchback throwing off his cloak and stepping forth into the light. These were often as entertaining to me as so many glasses in a magic lantern, and never took their shapes at my bidding, but seemed to force themselves upon me."

What the settlers had mostly encountered on crossing the Alleghenies was not menacing Indian tribes but trees. Much of Ohio and Indiana were blanketed by hardwood forests that had claimed the land after the last ice age, black, white, red, and yellow oaks, chestnuts, beeches, elms—red, white, and slippery. There were dogwood and poplar, white birch and black, wild crab and cherry, linden, ash, black walnut, box elder, sassafras, pecan, cottonwood, and a dozen more, their branches laden down by fist-thick vines that formed a vaulting canopy along the river bottoms. Stories were told of hollow sycamores that could hold thirty men. Once topped, roofed, and fitted with a door, such a trunk could serve as a smokehouse or a barn capable of sheltering several cows and horses.

Within a generation, perhaps two, almost all of it was gone. The ax was the major implement of destruction, but fire and wind also served to great advantage. Ian Frazier, the author of *Family*, writes that "the smell of settlements was the smell of burning woodpiles." When burning proved too slow or too dangerous, settlers took note of the prevailing winds and cut halfway through the trunks of several acres of trees, then waited for a major storm to roll in. When it hit, the trees on the windward side snapped in what sounded like a fusillade, setting off a massive chain reaction that leveled the rest, baring land and sky for the sowing of seed. In Frazier's words, "The settlers did not just attack the forest, they smote it."

Such was the pilgrim's progress of the nineteenth century, and it added greatly, if unknowingly, to the buildup of world carbon dioxide levels as the axmen marched west from New York in 1850, to Michigan in 1870, to Wisconsin in 1880, and to Paul Bunyan's Minnesota in 1890. Alex T. Wilson, a New

Zealand geochemist writing in *Nature*, notes that the pioneer explosion in North America was paralleled by rapid settlement and deforestation elsewhere, including South Africa, Australia, eastern Europe, and South America. By 1850 as much as half a billion tons of carbon from burning forests and other vegetation may have been injected into the atmosphere annually, perhaps equaling the grand total from the world's consumption of fossil fuels.

One of the few to grasp the magnitude of rampant deforestation was George Perkins Marsh, a granite-jawed Yankee possessed of a worldly eye. An attorney and master of classical languages (Marsh was fluent in twenty tongues by the age of thirty), he served as minister to Turkey under Zachary Taylor and later accepted a similar post in Italy at the behest of Abraham Lincoln. While abroad, Marsh made an extensive study of the geography, forests, and agricultural practices of Europe and the Middle East, providing the fledgling Smithsonian Institution with many specimens of native flora and fauna. In 1864 he published *Man and Nature, or Physical Geography as Modified by Human Action*, a prescient classic that would gain its author the title of first conservationist in the United States.

Marsh watched in dismay as pioneers claimed vast tracks of an essentially continuous forest, then converted them into grasslands with only scattered copses and groves. "The felling of the woods," he wrote, "has been attended with momentous consequences to the drainage of the soil, to the external configuration of its surface, and probably also to local climate." Not only "does an increasing population demand additional acres to grow the vegetables which feed it and its domestic animals, but the slovenly husbandry of the border settler soon exhausts the luxuriance of his first fields, and compels him to remove his household goods to a fresher soil."

In an insightful section titled "Instability of American Life," Marsh advanced a simple proposition: "We have felled forest enough, everywhere, in many districts far too much. Let us restore this one element of material life to its normal proportions, and devise means of maintaining the permanence of its relations to the fields, the meadows, and the pastures, to the rain and the dews of heaven, to the springs and rivulets with which it waters the earth. The establishment of an approximately fixed ratio between the two most broadly characterized distinctions of rural surface—woodland and ploughland—would involve a certain persistence of character in all the branches of industry, all the occupations and habits of life . . . without implying a rigidity that should exclude flexibility of accommodation to the many changes of external circumstance which human wisdom can neither prevent nor foresee."

His answer came in the form of a popular chapbook called *Paul Bunyan Comes West.* Dragging his pick behind him, the mythical lumberjack gouges out the Grand Canyon of the Colorado, then builds a tourist hotel from trees felled with his granite limbs. The top seven stories are mounted on hinges "so's they could be swung back for to let the moon go by."

~

While the saga of the westward movement was being played out, back in England Alfred Russel Wallace resigned his teaching position at the collegiate school in Leicester and in 1847 headed for an even more exotic frontier. Unlike Darwin, who was a man of wealth and position, Wallace had to scramble for a living by collecting insects and other biological specimens in the uncharted Amazon River basin, a quest that would eventually lead to his independent discovery of evolution.

It nearly led to his early demise as well. During Wallace's return voyage in 1852, his vessel caught fire and was destroyed, together with most of his live animals and valuable specimens. The naturalist and his fellow survivors spent ten anxious days in a lifeboat before being rescued, wasting and parched, but otherwise unharmed. A year later he published his *Travels on the Amazon and Rio Negro,* whose brisk sales enabled him to recoup a part of his financial loss.

He next sailed for the Malay Archipelago, as mysterious and unexplored as the Amazon. There he spent the next several years tramping Borneo, Bali, Java, Sumatra, Timor, Celebes, the Moluccas, and New Guinea, making him the most traveled naturalist in the world. In February of 1855, while Wallace was at Sarawak, in Borneo, he penned his first explicit essay on evolution; a second was composed three years later to the month at Ternate, in the Moluccas, as he was recovering from an attack of malaria. The latter paper, completed in just two evenings, was sent to Darwin by the next mail, launching a joint scientific revolution.

Wallace followed his first popular success with *The Malay Archipelago*, the most vivid chapter of which recounts in all too graphic detail his shooting of several orangutans, from youngsters and adolescents to mature adults, their skins and skeletons destined for museums and the trophy rooms of private collectors. Yet both of his works contain a vivid subtext—that of a man held in thrall by the vast and interminable rain forest girdling the equator. "Here [in Brazil] we may travel for weeks and months inland, in any direction, and find scarcely an acre of ground unoccupied by trees. The forests of no other part of the world are so extensive and unbroken as this. Those of Central Europe are trifling in comparison. . . . In North America alone is there anything approaching to it, where the whole country east of the Mississippi and about the Great Lakes is, or has been, an uninterrupted extent of woodland."

Though he was half a world away and out of communication for weeks, often months, on end, Wallace was at least

Alfred Russel Wallace
(Used by permission of Culver Pictures)

dimly aware of the fate overtaking North America's eastern woodlands. That such a transformation could occur in the Amazon was beyond imagining: "The banks of all these streams are clothed with virgin forests, containing timber-trees in inexhaustible quantities, and of such countless varieties that there seems no purpose for which wood is required but one of a fitting quality may be found." Canada, with its frozen rivers and benumbing temperatures, was of little consequence compared to thousands of miles of open water free of rapids and untouched by violent storms. Amazonian cedar could be cut and shipped to England more cheaply than white pine from Canada, whose gloomy forests had been permanently scarred by the woodman's ax, "while the treasures of this great and fertile country are still unknown."

During the century and more since Wallace put pen to paper, as much as 20 percent of the world's rain forests have been destroyed by human depredation, yet much of this still vast and canopied wilderness remains an enigma. Scientists have neither named nor cataloged most of the species that inhabit it, including many conspicuously large trees, "new" varieties of which are discovered every year. There are few jobs for taxonomists, and almost none in the third world, where their training is most needed. Thus, no one really knows, not even to within a factor of ten, how many kinds of plants, animals, insects, and microscopic organisms dwell within, much less how many have already vanished in the wake of fire and ax. Some say the total is 5 million, others claim twenty times that. Whatever the true number, it dwarfs the 1.4 to 1.7 million species that have been described by scientists since the time of Linnaeus in the eighteenth century.

Paradoxically, these massive ecosystems exist in what biologists term "wet deserts." Swathed in clouds of their own making, which limit the number of sunshine hours to between four and six per day, the many-layered trees prevent more than half of all the rainfall from ever reaching the forest floor. That which does falls on extremely poor soil whose few nutrients are quickly washed away into the streams and rivers in the valley floors. In evolutionary response, the dense root systems of the trees are concentrated in the uppermost soil layers, where they feed on the remaining 0.1 percent of the nutrients released by decomposing matter—the biological equivalent of the steady state universe. Yet a typical swath of forest measuring four miles square contains a staggering number of individual species: 1,500 flowering plants, 750 trees, 125 mammals, 150 butterflies, 100 reptiles, 60 amphibians, and an estimated 50,000 different insects, most of which hatch, mature, reproduce, and live out their brief lives in the canopy overhead.

First to fall victim to humankind were the rain forests of Asia. For centuries population growth in India has resulted in the continuous shrinkage of forestland. The need to obtain firewood for cooking became particularly acute following independence, and in many areas forests have simply ceased to exist. Official figures on the remaining amount of forested land—nearly one-fourth of the country's total—are highly misleading since much of it contains little more than worthless scrub. The compaction of the poorly protected soil has lowered its capacity to absorb groundwater, escalating the runoff of the monsoons and increasing the chances of severe flooding, siltation, and rapid erosion.

The most valuable tropical forest in Asia belongs to Indonesia and covers more than half the country. With almost 60 percent of it set aside for commercial logging in the early 1980s, trees were disappearing at a rate of 1.2 percent a year, aided by the introduction of floodlights that enable bulldozers and chain saws to operate round the clock. Thailand, Laos, Nepal, and Sri Lanka were losing just as much or more of their forests annually, with the Philippines and Vietnam not far behind.

Meanwhile, a short, well-knit tapper of rubber trees named Francisco Mendes Filho, better known to his friends as Chico Mendes, had undergone a crisis of conscience. A resident of Acre, an isolated state in far western Brazil, the affable and plainspoken laborer fought to establish a union to protect Earth's largest rain forest, endearing him to the members of the international environmental community. Mendes had also become an outspoken critic of the government's decision to carve a highway across his native province as part of a direct link to the Pacific, threatening the destruction of millions of trees and other organisms. He was flown to Washington and

London, where he spoke of his dream of establishing "extractive reserves," large tracts of land that could be used for harvesting renewable resources like the latex he and his fellow tappers collected in hundred-pound slabs.

On a late-December evening in 1988, Mendes, clad as usual in sandals, a T-shirt, and shorts, opened the back door of his dilapidated house in the village of Xapuri and began walking toward the outside toilet. He was met with a blast from a 20-gauge shotgun, fired point-blank. The funeral took place on Christmas Day, and a new cross soon appeared in the local cemetery, adding yet another name to the list of 1,000 slain unionists, priests, lawyers, laborers, and activists who had dared challenge the status quo.

The conditions leading up to Mendes's assassination had been spawned in the early 1960s, when Brazil's military rulers fostered the exploitation of the Amazon by building roads and offering generous tax and other incentives to industrialists and land speculators. Ranchers got into the act by purchasing and clearing hundreds of thousands of acres, transforming the forest into an artificial steppe for grazing without creating any significant employment in return. Once the roads were in, migration proceeded at a harrowing pace, spurred by a peasant class that has had to make do with a scant 2 percent of the farmland. Salvation, it was argued, lay in "opening up the jungle," a resource so vast that everyone from timber merchants to urban refugees would profit.

Thus the decimation began. The burning, which quickly achieved the status of a tradition, commences in September on the feast of St. Bartholomew, one of the Twelve Apostles, whose day of honor is said to bring especially good luck. Within hours huge areas of the forest are in conflagration, a blackened reminder of a martyr variously believed to have been flayed, beheaded, or crucified. As the quality and quantity of their crops decline, the thousands of squatters simply move on, much as did the generation George Marsh watched slash and burn its way across the U.S. frontier.

As for planted pasture, no tropical grass is well suited to trampling by cattle or sustaining itself during the dry season when the earth is laid bare. Dust, once unknown in Amazonia, has become commonplace, raising the specter of Oklahoma in the 1930s. After being jump-started by government, the opening up of the rain forest may simply be unstoppable.

~

That deforestation has added to the warming of Earth is a given among most atmospheric scientists. When the world's rain forests are set ablaze, the planet is subjected to double jeopardy. Burning trees give off carbon dioxide, and burned-out forests can no longer remove the gas by locking it up in their cells. Removal of the forest biomass and its replacement by pasture or crops means that at most 20 percent of the carbon content of the former woods is fixed in the new vegetation. The rest is oxidized, either by burning or decay, and enters the atmosphere as CO_2. Some of it disappears into the ocean sink; the rest remains in the global atmosphere to mingle with the billions of tons of CO_2 emitted annually by spent fossil fuels previously sequestered underground.

Beyond this, some climatologists hypothesize that the continued decimation of the rain forest will create a secondary greenhouse effect. As they see it, the reduction in rainwater will interrupt the continuous cycle of evaporation, which produces a natural cooling of the air. The result could be a local temperature increase of three to five degrees Celsius, well beyond that predicted for the tropics by global warming. And should a critical amount of the Amazon's biomass succumb to human intervention, any forest reserves would gradually expire from desiccation, thus completing the transition from verdant wilderness to a nascent Sahara on the march.

Ironically, environmentalists were making a concerted effort to focus international attention on the disappearing rain forests at the very time Joseph Farman's discovery of the ozone hole was being announced. They believed their prayers were at least partially answered in 1985, with the formation of the Tropical Forest Action Plan under the auspices of the United Nations. Promising both conservation and sustainable development, the plan proved ineffectual from the outset, in part because the targeted nations saw it as yet another aspect of neocolonialism.

At a meeting of the Amazon Pact in Quito, Ecuador, the host country's president, Rodrigo Borja, declared, "We will not permit external interference in Amazon affairs." At this same meeting, President Jose Sarney of Brazil complained that the developed world is principally to blame for the emission of greenhouse gases resulting in acid rain and the depletion of the ozone layer. Sarney's foreign minister had the figures to prove it, and cited statistics showing that the average American consumes energy at fifteen times the rate of the average Brazilian. Moreover, nobody was complaining about the damage that would likely result if China and India decided to exploit their massive coal reserves.

Officials were further rankled by the perceived condescension of self-righteous Yankees suddenly bitten by the environmental bug, including the newly rich makers of specialty ice cream and rock stars such as the Grateful Dead. The liner of the group's album *Deadicated,* whose royalties were earmarked for preservation of the rain forest, states: "Don't buy tropical wood products. Refuse to use disposable chopsticks (bring your own). Don't eat fast-food hamburgers. Two-thirds of [Central America's] rain forests have been cleared to raise cattle."

Like it or not, saving the tropical forests had rapidly become a global issue. What had taken millions of years to evolve was threatened with utter destruction in mere decades. In contrast to gold, platinum, or diamonds, these giant ecosystems no more seemed the property of individual nations than the seas, the sky, and the stars. Their annihilation would have an incisively devastating effect on all life on Earth, perhaps as profound as the return of the Pleistocene ice or the long dark winter ushered in by a careening asteroid 65 million years ago.

It was the force of this argument filtered through the Western media and vocal environmentalists that ultimately caused the government of Brazil to respond as it rarely had before. A burst of police activity resulted in the arrest of the father and son ranchers Darly and Darcy Alves da Silva, both of whom were charged and eventually convicted of the murder of Chico Mendes. The two men, who were each sentenced to nineteen years' imprisonment, had been purchasing chunks of the Amazon and clearing the forest for grazing. Subjected to constant harassment by Mendes, who had also learned of an outstanding arrest warrant for murder on the younger da Silva, they hired an assassin to silence their nemesis forever.

The conviction of Mendes's killers led to progress on another front. In the early 1990s, the Brazilian government, with the backing of the World Bank, established four extractive reserves open to rubber tappers and nut harvesters, most of whose families had moved into the forest early in the twentieth century, drawn by the great rubber boom that peaked before World War I. Now, in the words of the bank, forest dwellers would be able "to defend their homes and resource base against often violent encroachment" from ranchers and others.

The largest of these, the Chico Mendes Reserve, covers some 3,850 square miles, approximately half the size of Massachusetts. Significant as it is, the reserve constitutes less than half the amount of the Amazon lost to deforestation almost every year since 1980, a statistic of which the tourist industry is keenly aware. A frequently run ad in the *New Yorker* proclaims, "The AMAZON: Discover the Rainforest before it disappears!"

THE CLIMATIC FLYWHEEL
~

I've looked at clouds from both sides now
From up and down, and still somehow
It's cloud illusions I recall
I really don't know clouds at all.

—Joni Mitchell, "Both Sides, Now"

So halcyon was the late-summer weather of 1991 on the Tyrolean border between Austria and Italy that a German couple named Simon extended their climbing vacation another day. The two were at 10,500 feet and heading down toward their lodge after successfully negotiating the area's second-largest peak when they spotted the upper part of a body protruding from the ice. It lay facedown in a small depression and appeared to be naked.

Police, working under the watchful eyes of curious onlookers, made two unsuccessful attempts to free the body during the next few days. Since it had obviously been in place for some time, most assumed it to be that of another unlucky climber who had met with an accident, perhaps decades earlier.

Four days after its discovery, the corpse was finally freed by Dr. Rainer Henn, head of the Forensics Department at the nearby University of Innsbruck, who had been flown to the site by helicopter under a court order. The none too delicate extraction was performed with a ski pole and ice ax in front of television crews drawn by intriguing rumors that the body might actually be several centuries old.

In the typical body retrieved from a glacier, the moisture will have turned the tissues into a fatty, pliable substance known as "grave wax." Henn immediately noted that the corpse before him was dehydrated, like the mummies from ancient tombs, yet there were no clear signs that it might be that of a waylaid caller from another time.

As the body was about to be loaded aboard the helicopter, someone spied a flint knife with a wooden handle. The discovery prompted other artifact seekers to come forward, including one who had picked up an ax whose head was made of copper. After consulting with archaeologist Konrad Spindler, Henn and his colleague held a news conference the next day, where they announced that the body was that of a man who had perished 4,000 years ago. To verify this finding, several carbon-14 tests were conducted, the results of which extended the time line back another twelve to thirteen centuries. The Iceman, as he came to be popularly known, was some 5,300 years old, a relic of the late Stone Age or perhaps, as some would argue, of a lost Bronze Age.

Discovery of the Iceman
(Photograph by Paul Manny; used by permission of Gamma Liaison)

Graced by tattoos and crowned by a thatch of dark, wavy hair, the twenty-five- to forty-year-old pilgrim had likely committed the fatal mistake of lying down to rest out of exhaustion and then falling asleep. Huddled in a small trench, the body was freeze-dried by the frigid mountain air, then blanketed by successive layers of snow that gradually compacted, becoming glacial ice.

In the eyes of climatologists, Iceman's sudden deliverance resulted from the reversal of the very climatic conditions that had conspired to preserve him for five millennia. During the past several decades his alpine glacier, like others in mountainous regions all over the world, has gradually melted away, exposing underlying rock and gravel—an exciting prospect for the archaeologist. But for others the shrinking glaciers are an ominous harbinger of global warming triggered by human activity.

~

That message was being driven home on a number of other fronts. The sudden and frightening appearance of a continent-size ozone hole over the Antarctic in 1985 prompted delegates from forty-three nations, both rich and poor, to gather in Montreal only two years later at the goading of environmental scientists, who projected 130 million additional cases of skin cancer over the next century if nothing were done to stem the production of chlorofluorocarbons. Added to this was the potential for widespread genetic damage to crops and forests, as well as the boosting effect on global warming. In the words of Michael Oppenheimer, a scientist in the employ of the politically active Environmental Defense Fund, "Looking at the ozone hole is like staring at a picture where you watch the future of the human race go down a big black hole."

Opponents of the environmental movement accustomed to a sympathetic hearing from inside the Reagan administration argued, in all seriousness, that it would be better to have people rely on more sunglasses and suntan oil than to restrict the CFC industry. This time, however, they were rebuffed by Lee M. Thomas, administrator of the Environmental Protection Agency, who dismissed their lame palliatives out of hand. Exactly who was going to supply the poor of the world with hundreds of millions of pairs of sunglasses in addition to untold gallons of sunscreen?

Deputy Assistant Secretary of State Richard E. Benedick, vested with the rank of ambassador, led the U.S. delegation at the September gathering, the month of the year that had always marked Joe Farman's return to the ice with the British Antarctic Survey. Now, in the very midst of the Montreal negotiations, another expedition was at the bottom of the world taking the latest measurements of the ozone layer, while, far to the north, squatters were engaged in their annual rite of setting Amazonia ablaze. The first reports to reach the delegates confirmed a 40 percent reduction in the ozone level, triggering speculation that the Arctic might yet suffer a similar fate despite previous assumptions to the contrary.

The tenuous ground of international diplomacy suddenly firmed up. On September 16, 1987, agreement was reached on the Montreal Protocol, hailed by its twenty-four signatories as a "milestone in international cooperation to safeguard the environment." Its major provisions called for the freezing of CFC use at 1986 levels by 1989, and for slashing consumption by 50 percent of 1986 levels before the millennium. At the press conference held in conjunction with the signing, an exuberant Benedick noted that "for the first time in history the international community has initiated controls on the production of an economically valuable commodity before there was tangible evidence of damage."

Reaction to the agreement was not as polarized as had been anticipated, no doubt because the use of CFCs in aerosol sprays had been banned in the United States since 1978. While spokespeople for the five largest U.S. manufacturers of CFCs complained that their research costs would increase significantly, they were also reasonably confident that substitute compounds could be developed for use in certain industrial processes, although meeting the protocol target dates was problematic. Environmental groups, including the Natural Resources Defense Council, the World Resources Institute, and the Environmental Defense Fund, unanimously praised the document, but they were concerned that it did not go far enough in restricting emissions of the damaging chemicals.

Among the least impressed was Sherwood Rowland, who had been battling the chemical industry tooth and nail from that day, thirteen years earlier, when he and Mario Molina had published their paper in *Nature*. Who could say for certain,

Rowland mused, whether the signatories would keep their bargain, or whether the other nations of the world that had not signed the protocol would embrace the partial ban anytime soon? Meanwhile, more than 8 million tons of CFCs had been produced since 1974, 5 million of which were still on the loose, eating their way through the exquisitely delicate fabric of the stratosphere. Only time would tell if it was already too late to save the ozone layer in light of the damage that would continue for another century, no matter what anyone might do.

~

Nine months later, in stifling heat reminiscent of the mid-1930s, the esoteric contents of scientific journals were suddenly thrust into the spotlight. On June 23, 1988, with the temperature at a record 101°F on what was only the second day of summer in Washington, D.C., James E. Hansen, director of NASA's Goddard Institute of Space Studies and a widely known climatologist, was called to testify before the U. S. Senate Committee on Energy and Natural Resources. Spare and intense, Hansen pulled no punches: "The greenhouse effect," he warned, "has been detected and it is changing our climate now." Even more startling, Hansen was "99-percent confident" that current temperatures represent a "real warming trend" as opposed to natural variability. "We're loading the climate dice."

Senate committee chairman J. Bennett Johnston of Louisiana echoed Hansen's views: "We have only one planet," Johnston declared. "If we screw it up, we have no place to go. The greenhouse effect has ripened beyond a theory." Johnston's fellow Democrat, Arkansas senator Dale Bumpers, sounded a populist alarm: "What you have is all the economic interests pitted against our very survival." Bumpers went on to declare that Hansen's testimony "ought to be a cause of headlines in every newspaper in the country," which it was the next morning.

Conditions deteriorated as the summer of 1988 wore on, enveloping Hansen in the aura of a prophet. The interior of the United States sweltered through its worst drought in fifty years, while parts of the Southeast were drier than they had been in a century. Across much of the Great Plains and Midwest, only three months—June 1933, May 1934, and June 1936—had ever been drier than June 1988. Staring grimly at his parched and barren acres, an eighty-five-year-old Montana rancher sighed, "I've never seen it like this, not even in '34. My God, look at it. It's as bare as a dance floor."

Low rainfall was accompanied by extreme heat. Monthly record-high temperatures were set in thirteen cities in May, in sixty-nine in June, in thirty-seven in July, and in thirty-one in August. New York City suffered through more than forty consecutive days of abnormal heat and humidity. In Los Angeles, 400 overburdened electrical transformers blew up in a single day as temperatures peaked at 110°F in September. Even relatively cool Seattle broke 90°. Nationwide, 2,000 daily temperature records were set, while an estimated 10,000 "excess" deaths were linked to heat stress. Needless to say, the weather was the dominant topic of conversation, fueled by a spate of articles and television reports that stirred visions of the once and future Dust Bowl.

While crop yields declined by 30 to 40 percent and barges became stranded along the shrunken channel of the Mississippi, proving the river worthy of its nickname—the Big Muddy—Muscovites were fleeing a capital enduring its hottest summer of the century. Any body of water that promised relief, including polluted rivers and ponds, would do. By season's end the number of Russian swimmers who had drowned was more than double the previous high. And just as the experts on global warming had anticipated, other parts of the world were subjected to the horrors of water without end. In late August almost 80 percent of Bangladesh was inundated by a massive flood that left 25 million without shelter. This deadly invasion was followed by another in the form of cholera and dysentery epidemics. Once ravaged by a Noachian deluge every half century or so, this was Bangladesh's fifth such flood since 1980, each more devastating than the last.

Hansen's credibility received a further boost when it was determined that 1988 was the warmest year on record, and that the four warmest years prior to that one had also occurred in the 1980s, making it the warmest decade in 127 years, with two years still to go. When asked by a reporter from *Science News* about the scientific validity of his testimony,

Hansen refused to temporize: "There's no time at which you're 100 percent certain, but if we look at the record, I think it's beginning to get pretty darn clear that something is going on. And in my opinion it's time to say that."

~

Below a hill near a lake in the Adirondacks is the home of nature writer Bill McKibben. While Hansen was testifying on Capitol Hill, McKibben was completing a lengthy article on global warming for the *New Yorker*. A year later, this spiritual descendant of George Perkins Marsh and John Burroughs expanded the piece into a slender volume, which became the best-seller *The End of Nature*. Believing that we have already entered the greenhouse world, the author fears for the generation about to inherit the planet.

For all of human history and long before, McKibben writes, the wind and weather, clouds and rivers, forests and oceans were part of the natural order against which nothing could prevail. But by the late twentieth century humankind had succeeded in altering the most basic forces around it through the massive consumption of fossil fuels: "We have changed the atmosphere, and all that will change the weather. The temperature and rainfall are no longer to be entirely the work of some separate, uncivilizable force, but instead in part a product of our habits, our economies, our way of life. . . . The world outdoors will mean much the same thing as the world indoors, the hill the same thing as a house."

This transformation—as profound in its own way as the Neolithic revolution—was the result of no malign intent, for who ever dreamed of wrecking nature by placing humanity's thumbprint on every raindrop, its signature on every breeze that stirs. "But the *meaning* of the wind, the sun, the rain—of nature—has already changed. Yes, the wind still blows—but no longer from some other sphere, some inhuman place." A child born today will never know a natural summer or autumn or winter or spring. "We have deprived nature of its independence, and that is fatal to its meaning. Nature's independence *is* its meaning; without it there is nothing but us."

McKibben admits to having seriously considered joining the ranks of the radical monkey wrenchers, the legal consequences of disabling construction machinery and sabotaging logging operations be damned. Meanwhile, he and his wife have begun "to prune and snip our desires." They ride their bikes instead of taking long drives; they opt for thermal-pane windows rather than a hot tub in the backyard; they heat with wood and try to keep the house at 55°F.

On a more melancholy note, the couple has been unable to resolve the question of whether to bring a child into a world where unchecked global warming could literally rip civilization apart. And this, McKibben muses, is because they are unlucky enough to live at a time when an invisible gas is approaching an intolerable level, and the rain forest girdling the planet is rapidly turning into a blackened waste that could endure for millennia—a fugue to the end of time.

~

Having assumed a high profile, James Hansen became a hero to many, while others, including a number of his fellow scientists, accused him of jumping the gun. In the spring of 1989, a major workshop on climate change was held in the university town of Amherst, Massachusetts. Among those attending was Michael Schlesinger, a climate researcher at Oregon State University, who was privy to the same data as Hansen. "Taken together, his [Hansen's] statements have given people the feeling that the greenhouse effect has been detected with certitude," Schlesinger charged, but "our current understanding does not support that. Confidence in detection is now down near zero."

An equally skeptical stance was taken by a bearded and prematurely balding Richard Lindzen, who, though characterized by a writer for the journal *Science* as the "commander-in-chief of the counterrevolutionary forces," looked strangely like the young Karl Marx. Though his troops, at least the ones willing to speak out, were estimated to number less than a dozen and in most cases had not specialized in greenhouse research, they were emboldened by the credentials of their general—a Harvard man, Sloan Professor of Meteorology at the Massachusetts Institute of Technology, and member of the National Academy of Sciences.

Lindzen rejected the increasingly popular view advanced by the rumpled geochemist Wallace S. Broecker that "the climate system is an angry beast and we are poking it with sticks." If we continue to do so, Broecker warns, it threatens to turn on us like Blake's immortal Tiger: "What the hand dare seize the fire?"

According to Lindzen, time scales of a few centuries or less mean little when it comes to global warming or cooling. Furthermore, the computer models employed by Hansen and others to support their doomsday predictions are so rife with uncertainty as to be useless. Greenhouse gases will inevitably increase in the future and could conceivably double at some point in the next century. But the atmosphere, ever in quest of stability, will trigger its own natural immune response in the form of more heat-deflecting clouds, warming at most by a few tenths of a degree; in effect, next to nothing.

It was to Lindzen, not Hansen and his allies, that President George Bush and his White House chief of staff, John Sununu, were listening. In September of 1989, Lindzen, together with Jerome Namias of Scripps and Reginald Newell of MIT, coauthored a letter to the president in support of a thirty-five-page document issued by the George C. Marshall Institute, a conservative Washington think tank. Titled "Scientific Perspectives on the Greenhouse Problem," the report was drafted by three prominent scientists, William A. Nierenberg, director emeritus of Scripps; Robert Jastrow, astronomer and former director of the Goddard Institute of Space Studies; and Frederick Seitz, president emeritus of Rockefeller University and past president of the National Academy of Sciences. This same trio had vigorously supported President Reagan's Strategic Defense Initiative a few years earlier, whereby incoming enemy missiles would supposedly be knocked from the sky by satellite-mounted lasers commanded and triggered by computers.

Summing up the many unknowns surrounding the greenhouse models and predictions, the authors, none of whom are experts on global warming, conclude that it is too early to take any action to reduce greenhouse gases. The modest temperature rise of about one degree Fahrenheit during the twentieth century cannot, with any certainty, be attributed to fossil fuel emissions. They further argue that decreased solar activity in the next century will result in a cooling trend likely to offset the current warming.

Believing that this nonrefereed document was at least partly behind the White House's temporizing on climate change, greenhouse experts dismissed it as "biased" and "misleading," an affront to those who had authored extensive and carefully researched scientific monographs during the 1980s. Steve Schneider of the National Center for Atmosphere Research in Boulder, Colorado, denounced the report as a political document, and he did not pull any punches when it came to Lindzen's views. "Does he have a calculation, or is his brain better than our models? You can't just sit there and build a model of one sector of the atmosphere, then extrapolate to the globe. That's why you build models."

~

Clouds—the ethereal bedrock of model building. We can still recall them from the pages of our junior high science texts, having memorized their outlines for the weekly quiz. Cirrus: high-altitude patches of white, featherlike plumes; cumulus: dense, white puffy globs against which our brightly colored kites swam like sky fish into the spring wind; stratus: melancholy layers of low-hanging slate, resembling a horizontal fog bank. The hybrid forms were soon forgotten, except, of course, the Jovian cumulonimbus, which pile up against the horizon like dark, angry mountains. Depending on the time of year, they were generally good for rain, snow, hail, or a walloping thunderstorm, what my grandfather used to call a "gully washer."

We were also taught that clouds are formed by droplets of water or, at higher altitudes, of ice crystals suspended in the air. When the droplets are large enough to overcome the upward-moving currents, they fall to Earth as drizzle or rain. However, I do not remember being taught that a classification of cloud forms was first undertaken in 1801 by Jean-Baptiste Lamarck, the great French naturalist mostly remembered for his theory of acquired characteristics. Or that two years later an English scientist named Luke Howard devised a classification scheme designating three primary cloud types and another ten or so compound forms. (Adopted, with modifications, by the International Meteorological Commission in 1929, Howard's system remains in use today.) More important still, we did not learn that the fate of Earth as we know it is dependent on myriad water droplets infusing the planetary atmosphere.

Climatologists estimate that at any given moment clouds

cover at least half the planet and perhaps as much as 10 percent more. Acting as a giant umbrella, they bounce an estimated 20 percent of the Sun's light back into space before it can warm Earth's surface by another twenty-two degrees Fahrenheit. This reflective capacity is what scientists term albedo. Light-reflecting surfaces such as ice and snow also have a high albedo; dark, light-absorbing surfaces such as large bodies of water have a low albedo. Because clouds dampen the Sun's warming effect, they are said to produce a negative feedback.

Inasmuch as clouds are composed of water vapor, far and away the most abundant of the greenhouse gases, they also exert a contrary effect by absorbing some of the infrared radiation emitted by Earth. This heat flows back downward, warming the planet and causing a positive feedback by amplifying the change. Anyone who has felt the cooling effect of a passing cloud on a hot day has witnessed a negative feedback, while anyone who has walked the city streets on a warm, cloudy night has felt the blanketing effects of a positive feedback. In the words of the song, "You've looked at clouds from both sides now."

So have puzzled climatologists, who by the late 1980s had long been faced with the conundrum of determining which of the two feedbacks is dominant—warming or cooling. If, as they suspected, the massive quantities of carbon dioxide being released into the atmosphere are warming the planet, sea surface temperatures should also be on the rise. With the aid of satellite imaging, such a temperature increase was, in fact, detected in the waters of the world. Furthermore, warmed oceans have a higher evaporation rate, adding more water vapor to the atmosphere. The ultimate result of this accelerated hydrological cycle is the formation of even more clouds.

According to Lindzen and scientists of a like turn of mind, additional cloud cover will boost negative feedback by reflecting more light and cooling the planet, thus short-circuiting the glum scenario advanced by environmentalists. Yet many of Lindzen's peers are skeptical. Their research indicates that a substantial increase in clouds will contribute to a warming of the air and sea, bringing about a number of positive feedbacks such as the melting of the polar ice.

In the Northern Hemisphere, where the bulk of the green-house gases are produced, the snow line has been retreating for the past century, as have the glaciers, whose lingering presence still proclaims us citizens of the last ice age. Where snow and ice, with their high albedo, once reflected the Sun's rays back into space, recently exposed mountains, rock, and tundra absorb more solar heat, leading to greater melting and further amplifying the change.

The warming tundra releases methane, one of the most potent of the greenhouse gases, while the melting ice raises the level of the surrounding sea. Not only is there more water, the sea itself expands when heated, requiring greater space than when it was colder. Although estimates vary, many scientists agree that if global warming continues unchecked, the world's sea level will be six inches higher by the year 2030 than it is today; in 2100, little more than a century from now, it will be some twenty inches higher than at present. Should the glaciers of Antarctica and Greenland enter into a suicide pact, things could get much worse, reason enough to keep a wary eye on both poles.

Not only do we have a lot riding on clouds, the shape, color, and location of these ephemera are of deep concern to experts who believe that we have become directly involved in their creation. For all their might and grandeur, towering cumulonimbus clouds are perhaps the least important when it comes to climate change on a global scale. Stratus clouds, which are gray, dense, and low-flying, have a net cooling effect because their albedo is relatively high. Conversely, wispy, high-flying cirrus, the "mare's tails" of our youth, are semitransparent to incoming sunlight but block infrared radiation emitted by Earth, contributing to the greenhouse effect. Beyond this, much remains unknown.

~

Like Darwin, who once called the mechanism by which a new species comes into the world "that mystery of mysteries," climatologists are confronted with their own mysteries of mysteries, the veiled dynamics of the climatic flywheel. In hopes of unraveling this enigma, they have turned to computers of the highest order to solve complex equations representing clouds, wind, temperature, ocean currents, albedo, sunlight, and greenhouse gases. Seated, like technological gods, before a magic keyboard, they brighten or dim the Sun with a scarcely audible click, speed up or retard Earth's rotation, create or delete clouds of every variety, increase rain, initiate an arctic thaw, sear croplands, multiply the waters of the seas, spin Milankovitch cycles, darken the continents, and double the amount of carbon dioxide in the atmosphere.

Beginning in the late 1960s and early 1970s, planet modeling was a fairly primitive venture, usually undertaken by one or two investigators. Their templates were known as zero-dimensional models because they did not resolve any of the latitudinal, longitudinal, or vertical patterns of the climate system. Soon, however, modelers were running a vertical line from Earth's surface out to the far edge of the atmosphere. At several points along the line, equations simulated the transfer of solar energy to Earth, the transfer of infrared energy from Earth to the atmosphere, and the upward movement of air via convection. The basic physics of differing cloud formations was factored in along with the concentrations of gases. Despite their one-dimensional character, these models were quite useful in simulating the effect of rising carbon dioxide levels on the atmosphere.

The rapid advance of technology led to the development of the first three-dimensional models in the mid-1970s, far more sophisticated versions of which currently hold sway. They are based on labyrinthine computer programs devised by teams of climatologists composed of some of the best minds in applied mathematics and atmospheric physics. In contrast to the early days, most models used in greenhouse research are referred to by an organization rather than the scientists who built the model: the Geophysical Fluid Dynamics Laboratory (GFDL) in New Jersey, the National Center for Atmospheric Research (NCAR) in Colorado, the Goddard Institute for Space Studies (GISS) in New York, and so forth.

Not only do these models require human intelligence of the first order, they can be run only on the world's largest and fastest computers. This costly enterprise has created a certain degree of dissension among researchers who are not a part of "big science." They have seen their financial support shrink, while funding for the leading modeling groups, who are often perceived as blinkered captives of their own technology, continues apace. The modelers counter that it is their critics who are biased, for they refuse to seriously consider findings that don't square with their largely conservative views.

The most sophisticated of the computer programs is the General Circulation Model, or GCM. Encompassing the planet, it simulates the climate by creating realistic boundaries for the oceans, continents, mountains, and cryosphere, the global aggregate of glacial ice and snowpack. Added to these are the intricate motions of the atmosphere and the on-again, off-again action of vast, deep ocean currents—the "great conveyor belt," as it is called by its explicator, Wallace Broecker, who, incidentally, performs his mathematical calculations the old-fashioned way, with pad and pencil. Mightier than a hundred Amazon Rivers, this snaking giant carries fully 30 percent as much heat to the North Atlantic as comes from the Sun. When it goes into hibernation temperatures plunge; when roused, a rapid warming ensues. Broecker postulates that these cycles can occur within as little as a decade for reasons that are still unknown, a theory backed by core samples of arctic ice.

Unlike the computer models used in weather prediction, which are programmed to run for a few days or a few weeks into the future, GCMs are constructed to simulate climatic evolution over months, years, decades, or even centuries, a tall order indeed. A fully developed model should deal with all of the physical, chemical, and biological processes that occur in nature, plus the interactions among the components of the climate system. And this must be done in a manner that reflects the ongoing dynamics of a global climate machine that is forever in flux.

Despite having made tremendous strides in recent years, climate modelers are the first to admit that theirs is not an exact science. According to Kevin E. Trenberth of NCAR, the latest atmosphere-ocean-sea-ice models are providing some "very good simulations of average climate conditions and their evolution with the seasons." Nevertheless, "climate models do contain obvious errors in some features when compared with observations." When such an error occurs, modelers are wont to introduce an artificial fix known as "flux adjustment," which is designed to keep the simulated climate in sync with that of the real world. Scientists are taught to abhor the fudge factor, and though the flux adjustment cannot be equated with outright cheating, climatologists would much prefer to dispense with it altogether. For this reason a great deal of research is focused on eliminating the need for this crutch.

Among the many other difficulties facing modelers is that of spatial resolution. Even the most advanced computer simulation provides feedback in the broad brush strokes of a van Gogh rather than the controlled dots of the pointillistic Seurat. Many climatic phenomena occur in areas smaller than those covered by the computer's grid system, each square of which represents hundreds of kilometers on a side. The majority of clouds, for example, cover only a few kilometers rather than the few hundred necessary for detection. Even massive thunderstorms and hurricanes are too small to be well resolved on a GCM's grid, so "they pass through," in the words of meteorologist Richard J. Sommerville, "the way a small bug can pass through a coarse window screen." The cumulative effects of this global trespass can badly skew any model, requiring that flux adjustment, like Ptolemy's epicycles, be called upon to save the day.

As with circulation features, the simulation of temperatures can slip through some very wide cracks. On a global scale the temperatures of the planet are quite well represented and are becoming more accurate all the time. The trouble comes with temperature variations on a regional scale, which are regularly subject to significant computer error. Moreover, no computer in existence is fast enough to calculate climatic variables everywhere on the planet and in the atmosphere in a reasonable time, nor will there ever be one capable of handling that many equations.

The one unknown that haunts modelers more than any other is the uncertainty of clouds. Their variability is enormous: Their longevity runs the gamut from microseconds to weeks, their size from invisible submillimeter dwarfs to gray giants blanketing thousands of kilometers of sky. Calculating their influence on incoming solar and outgoing infrared radiation constitutes a modeler's nightmare.

It is hardly surprising that clouds are treated differently in each of the major climate models, particularly in terms of their sensitivity to carbon dioxide. For example, one model renders them three times as sensitive to the greenhouse gas as another. Once the computer programs have been put into operation, the levels of CO_2 are gradually raised until they double, just as they are expected to do in the next fifty to one hundred years. This exercise is called "forcing" and parallels the pressures to which nature is subjected on a daily basis.

As expected, the model in which cloud cover changes least predicts a warming of about 1.5°C (2.7°F), while the one in which clouds change most dramatically projects a temperature increase of about 4.5°C (8.1°F), sufficient to stir visions of Revelation's Four Horsemen.

As matters stand, GCMs are accurate to within 10 percent of the magnitude of natural phenomena such as winds, rain, temperature, radiations, and cloud cover, a margin of error too great for making a reliable forecast about global warming, particularly one stretching far into the future. At the same time, the most advanced GCMs all project a continuing rise in global temperatures, one that has been accompanied by heightened CO_2 emissions over the course of the last century. These models also agree that when CO_2 is doubled, more high clouds will likely be created, and it is wispy cirrus that block the least amount of sunlight. Stratus, on the other hand, with their high albedo and close proximity to Earth, will be less prevalent than today, their decline likely marked by a reduction in the overall cooling effect and a rise in solar heating through positive feedback.

The ongoing increase in the surface temperature of Earth has been traced back as far as the 1880s, near the time of Darwin's death. Its ascent has been marked by occasional hesitations and one prolonged reversal lasting thirty years. Meanwhile, the rise of CO_2 has been linear, as has that of other greenhouse gases billowing invisibly skyward from our fires, smokestacks, and tailpipes. Darwin never resolved his grand mystery of mysteries—the nagging riddle of species creation—and went to his grave peering at nature through a glass darkly. As an unsung Austrian monk had already discovered, the answer Darwin sought lay hidden in the cells of mutating peas.

Many would argue that global warming itself is a muta-
tion, a Hydra-headed monster loosed on an unsuspecting
world. Yet our forecasts are uncertain, and our crystal ball is
swirling with myriad contingencies. It is the scientist who
comes face-to-face with this dizzying blur each day, and the
supercomputer that may offer the greatest hope of fathoming
the climatic flywheel.

Recently an NCAR model incorporating a host of refine-
ments was run for 300 simulation years and generated a stable
climate quite close to nature. What is more, its programmers
were able to do away with flux adjustments, which have been
employed for so long that they were verging on the re-
spectable. Researchers were further heartened when statistical
analysis determined that this correspondence was highly unlikely to have arisen by chance. According to Trenberth, when
greenhouse gases were factored in, the projected rates of climatic change "exceed anything seen in nature in the past
10,000 years." These included rising sea levels, water vapor increases over tropical seas, elevated temperatures for an already
feverish planet, and more frequent bouts of extreme weather
across North America, what climatologist Benjamin Santer,
chief author of a 1995 United Nations report, calls the "greenhouse fingerprint."

The key to this promising breakthrough was finding a
better method of incorporating the effects of ocean eddies,
swirling pools of water up to 200 kilometers across that spin
off strong currents, moving heat around the globe. Like
clouds, they are difficult to incorporate into climate simulations because they slip, undetected, through the models'
coarse grids. While the revamped GCM doesn't possess a finer
mesh, it includes a new "parameterization" that passes the
effects of these unseen eddies onto the larger-model scales in a
more realistic manner than before.

More than halfway across the country from Boulder, at
Princeton University, a team of researchers that includes
Sukuro Manabe, a longtime student of global warming, has
been involved in another kind of modeling. Employing a
1,000-year time series of global temperatures derived from a
mathematical model of the ocean-atmosphere-land system,
they found that "no temperature change as large as 0.5°C per

century is sustained for more than a few decades." Yet for much of the past century the air temperature of Earth has been increasing at this very rate. If the model is correct, this continuing trend is not a natural feature of the interaction between the oceans and the atmosphere. "Instead," write Manabe and his associates in *Nature*, "it may have been induced by a sustained change in the thermal forcing, such as that resulting from changes in atmospheric greenhouse gas concentrations and aerosol loading."

These developments notwithstanding, the most outspoken critics of modeling have refused to give ground. When interviewed by a reporter for the *New York Times*, a hard-bitten Richard Lindzen reasserted his position that "models are so flawed as to be no more reliable than a Ouija board." When asked to comment, the normally soft-spoken Santer exclaimed, "That's garbage!"

"I think models are credible tools and the only tools we have to define what sort of greenhouse signals to look for," said Santer. If he is right, the human imprint on climate should emerge more clearly in the next few years as computers become ever faster and more powerful, the modern equivalents of the Delphic Oracle. If he is wrong, the Ouija board, however suspect, has been known to tell the truth every now and then.

16

CASSANDRA'S LISTENERS

~

In nature's infinite book of secrecy
a little I can read.

—William Shakespeare, *Antony and Cleopatra*

The buzzing on the convention center floor grew louder as the long-anticipated moment drew near. For many assembled in Atlanta on this January night in 1990, it would be their one and only chance to see a true scientific hero in action. There was a momentary hush when the slender, some thought gaunt, figure entered the large room, escorted by officers of the American Association of Physics Teachers. Then the audience burst into spontaneous applause and was rewarded by the familiar smile they had often seen on the *Tonight* show with Johnny Carson. It really was Carl Sagan, and he had come to receive the Oersted Medal, the association's highest honor.

Having shouldered the burden of publicly facing down the peddlers of pseudoscience and superstition, the irrepressible warrior chose the controversial topic of global warming for his acceptance speech. But first he reminded the assembly, composed largely of high school teachers, that the phrase "greenhouse effect" is a misnomer. Greenhouses work by preventing convective cooling, the same principle that applies to the interior of an automobile when heated by sunlight on a summer's day. Yet he was forced to admit that the "phrase is so widespread in atmospheric physics that we are stuck with it."

Then, as Earth's image was projected onto an oversize screen, Sagan asked his audience to contemplate the so-called ocean of air surrounding the planet. The thickness of the atmosphere, including every molecule and atom involved in the greenhouse effect, he noted, is but 0.1 percent of our planet's diameter. When the high stratosphere is included, the atmosphere is still less than 1 percent of Earth's diameter. "'Ocean' sounds massive, imperturbable. But the thickness of the air, compared to the size of the Earth, is something like the thickness of a coat of shellac on a schoolroom globe. Many astronauts have reported seeing that delicate, thin, blue aura at the horizon of the daylit hemisphere and immediately, unbidden, began contemplating its fragility and vulnerability. They have reason to worry."

The audience was with him now, as the slide of Earth was replaced by another charting global mean temperatures from 1900 to 1988. "You can see substantial wiggles, noise, in the global climate signal. But there also seems to be a clear upward trend. The five hottest years in the entire 20th century occur in the decade of the 1980s." Not wanting to alienate the uncommitted by appearing to wax polemical, the astronomer chose his next words with great care: "This is the sort of evidence that has led at least some scientists to conclude that the signature of the increasing greenhouse effect is already here—not just something calculated for the 21st century, but here now." The letters comprising that signature—some bold, some subtle—had been multiplying with each passing year and would continue to do so throughout the 1990s.

~

In early December of 1833, while the *Beagle* was several miles off the shores of northern Patagonia, the ship and crew were suddenly engulfed in a great cloud of butterflies. Even with the aid of a telescope, Darwin could not see a space free from the insects. "The seamen cried out, 'it was snowing butterflies,' and such in fact was the appearance." Because the day had been fair and calm, Darwin concluded that the insects had not been blown out to sea by an ill wind, but that they had voluntarily taken flight.

In 1992 Camille Parmesan, a graduate student at the University of Texas at Austin, climbed into a "ratty old" 1977 four-

Camille Parmesan
(Photograph by Joanna Pinneo)

wheel-drive vehicle and began a Darwinian odyssey that would end, four years and 45,000 miles later, with the publication of a groundbreaking paper in *Nature.* Her quest involved Edith's checkerspot (*Euphydryas editha*), a small winsome butterfly with a range from the rocky hills of Mexico to the alpine terraces of British Columbia.

In contrast to its Patagonian cousins, one of the most notable characteristics of Edith's checkerspot is its aversion to roam. An entire population often confines itself for decades to a piece of land no more that a few hundred yards square. After personally surveying more than a hundred populations of the species from Mexico to Canada, then correlating her findings with prior studies, Parmesan discovered an interesting pattern of local extinctions and colonizations. The southern populations native to Mexico were four times more likely to become extinct as those in Canada. She also found that mountain populations at elevations above 8,000 feet were considerably more robust than those living at lower altitudes.

Though experts have long known that organisms shift their ranges in response to climate change, direct evidence of such a reaction to the warming of the last century had been lacking. "What my study shows," Parmesan, now a biologist at the University of California, Santa Barbara, related in an interview with William K. Stevens of the *New York Times*, "is that we're apparently already seeing the effect, even though this is a small warming—[an] effect larger than I predicted."

Her conclusions made perfect sense to other experts on lepidoptera. Peter Kareiva, an ecologist at the University of Washington in Seattle, called it a very important paper: "People have hypothesized a good deal about how species are going to respond to climate change, but we don't have any good data prior to hers." Paul R. Ehrlich of Stanford, while internationally known for his population studies, is also an expert observer of Edith's checkerspot. Ehrlich was quick to embrace Parmesan's findings, noting that even a small change in the microclimate can disrupt "the very complex relationship between [this species] and its food plants," resulting in local extinctions.

Far removed from the range of Edith's checkerspot, flat-bottomed cumulus clouds drift lazily on light trade winds above a sparkling Caribbean Sea. It is late summer in the Virgin Islands, and there is little to suggest that something might be amiss other than the reports of swimmers that the waters near the shallow coral reefs seem unusually warm. But below the ocean's surface things are markedly different. Divers have become reluctant witnesses to an unsettling transformation.

The normally pink, green, and golden-brown corals have turned a dazzling white, making them visible from a much greater distance than usual. The phenomenon, which is lethal, is not confined to the Virgin Islands. Observers at various locations in the Caribbean report similar bleaching, as do others living in the Florida Keys and along the coasts of Panama and Costa Rica, where coral mortality has reached 70 to 90 percent. Bad as it is, the reefs of the Galápagos Islands are suffering worse; more than 95 percent of the coral have died.

Though weighing tons and appearing to be architectural in structure, corals are composed of myriad colonized creatures, called polyps. Each tiny animal resembles a hollow cylinder, closed at the base and connected to its neighbors by the gut cavity. Its oral opening is surrounded by one or more rings of tentacles, while its soft external tissue, like that of the sea anemone, overlies a hard structure of calcium carbonate. Within its transparent cells live symbiotic algae, which help to nourish the polyp by producing carbon compounds through photosynthesis. Other creatures also live in a mutually dependent relationship with the coral, their many colors vivified by the refracted sunlight.

When coral bleaches, as it does once the water temperature rises to 32°C, the delicate balance among the symbionts is upset. The algae disappear, leaving a white calcium carbonate skeleton, like the bleached bones of a desiccated steer. This was first noticed on a large scale in the early 1980s, following the outbreak of an unusually fierce El Niño in the Pacific. Coastal water temperatures rose several degrees Fahrenheit in some areas before returning to their norms. Then, in 1986, while El Niño was safely in hibernation, a rise in the number of reports of coral bleaching forced scientists to take a closer look at the phenomenon.

Several factors can cause bleaching, including increased ultraviolet radiation, excessive shade, disease, sedimentation, pollution, and changes in salinity. Yet none of these is as compelling as the theory that global warming is the primary culprit. Indeed, some scientists consider the death of the delicate corals to be the equivalent of the canary in the coal mine, the early warning of a rise in the ocean temperatures of the world.

Research undertaken at the University of Panama has linked coral mortality with high temperatures in a series of controlled experiments that mirror the effects of El Niño. Coral in saltwater tanks take the same amount of time to die at 32°C as in the ocean, a strong indication that laboratory conditions have replicated those in nature. Still, evidence for the 1986 warming and that of subsequent years is not as definitive. While bleaching appears to be associated with local temperature increases, limited knowledge regarding the physiological response of corals to stress and the inadequacy of seawater temperature records have so far made it impossible to know for certain. What scientists do know is that an increase in the water temperature of a degree or two Celsius over the next half century could spell disaster for the tropical latitudes.

Positive feedbacks, such as those associated with alterations in the range of a butterfly or the bleaching of a protective coral reef, are not the only evidence of global warming. In a remote part of northwest Tasmania, an island state of Australia, scientists have located a stand of Huon pines whose trunks bear the imprint of an ongoing temperature rise. The Australian-U.S. team has established a correlation between expanding tree ring growth and the temperature record since 1900, a period of increasing carbon dioxide emissions. The evidence is even more pronounced since 1965, when the tree ring index jumped 50 percent, and the temperature rose from just under 15°C to just above 16°C. Conversely, the index reflects a cold snap in the early 1900s, when pack ice in Antarctica expanded and icebergs drifted north into the normally temperate latitudes of the Tasman Sea.

The Huon pines under study are located at elevations twice as high as those inhabited by the other 95 percent of the species. Because of their altitude, the trees grow only in summer and are highly sensitive to temperature fluctuations. Furthermore, the site has not been subjected to unnatural influences such as acid rain, which stunts tree growth in the Northern Hemisphere. So remote and untouched is the region that the pines were only discovered in 1989 by an employee of the Tasmania Forestry Commission.

Huon pines are known to survive for up to 700 years, but other species of pines in the area live two and even three times as long. The data collected on each species indicate that there has been no episode in the past millennium to match the wide ring growth of the late twentieth century, including the Medieval Warm Period of the tenth through twelfth centuries. As important as the rise in temperature has been the growing concentration of carbon dioxide in the atmosphere. Experts believe the gas is working like an invisible fertilizer, a negative feedback that can be counted as one of the few beneficial effects of industrial pollution. Examination of the tree rings from other high-altitude pines in California confirms laboratory predictions that increasing CO_2 is fostering plant growth and may, in fact, be stimulating the replication of Earth's biomass. This is good news when one considers that locked inside the branches, trunk, and leaves of the average tree is a ton of CO_2.

Well above the Arctic Circle, near the base of the rough-hewn Brooks Range that separates Alaska's North Slope from the rest of the world, lies Toolik Lake. Named for the yellow-billed loon, or *toolik* in Eskimo, it has steel blue waters set against a vast treeless expanse of arctic tundra, seemingly frozen to the very core of the world. Each summer scientists and their graduate students come to Toolik Lake via the Dalton Highway that runs from Fairbanks to the oil fields of Prudhoe Bay. They poke, prod, and take the measure of this massive carbon sink.

Aided by funding from the National Science Foundation, the researchers believe they have chosen one of the best places on Earth to study a pristine ecology that may be on the brink of major upheaval. Predictions are that arctic temperatures could rise by as much as six degrees Celsius in summer and twelve degrees Celsius in winter during the coming decades. Should such a drastic warming take place, millions of acres of permafrost would thaw, transforming the tundra into a suppurating morass impenetrable by man or beast. At the same time, an unknown portion of the estimated 180 billion metric tons of carbon locked in the soil would be released into the atmosphere, altering the surrounding ecosystem and adding significantly to the world's climatic woes.

To date, no one can say for certain whether a change of this magnitude will occur, but scientists may have a shorter time to gather their data than was previously thought. Between 1979 and 1994, the average surface temperature of Toolik Lake during the summer rose by nearly three degrees Fahrenheit. The tundra, meanwhile, has begun to release carbon dioxide as warmer air penetrates its surface and aerates the upper soil levels. Making the most of their one ally—the midnight Sun—field-workers in this isolated and mosquito-ridden land labor round-the-clock seven days a week for three exhausting months. They then clean and batten down their tundra dacha, load up the vans supplied by the University of Alaska, and head south to civilization, leaving the bleak landscape to the arctic fox and roaming polar bear.

In the center of Greenland, at virtually the same latitude as Toolik Lake but hundreds of miles to the east, lies Summit Camp. There a small group of scientists is preparing for the rigors of spending a winter on the ice cap. The quartet occupies a building measuring thirty-six feet by thirty-two feet and called the Green House (for its color rather than its location). They are but four among dozens of scientists gathering ice core samples and other data that will help researchers better understand the past 100,000 years of Earth's climatic history.

In the Arctic, unlike the Antarctic, there is no topography, only snow, sky, and horizon as far as the eye can see. No living thing stirs. Dry as desert sand, the air is bitterly cold and bears no smells. During storms, the snow shoots sideways in the wind, mounting up in drifts the size of boxcars. Polaris, the North Star, is almost directly overhead, while the aurora borealis, ethereal and awesome, draws the spellbound scientists outside though the temperature is –40°F.

It is not to contemplate nature's tremendous power and raw beauty that these scholars have come. Aided by the National Science Foundation, they seek samples of carbon and other elements. Between late August and the beginning of April, 1,300 cylinders of ice will be collected and transported back to laboratories in the United States. Eventually, these silver-blue cores may provide further evidence that the greenhouse effect is indeed a global phenomenon.

Scientific concerns about the opposite end of the *axis mundi* are no less compelling. Some 130,000 years ago, during the last interglacial period, the world's oceans were about six meters higher than at present. One hypothesis holds that this rise in sea level was caused by the collapse of a massive antarctic ice sheet. Unlike the eastern ice sheets, which rest like a carapace on solid ground, the western ice shelf extends into the Ross Sea, with no undergirding save the restless waters below. It daily calves icebergs the size of football fields, but every once in a while the fractured ice parts in sections hundreds, and on rare occasions thousands, of miles square.

Scientists agree that if this gigantic shelf were to break away, as may have happened thousands of years ago, the seas would again rise between five and six meters, dwarfing the projected figure of fifty centimeters over the next century. In 1996 Stanley Jacobs, of Columbia University's Lamont-Doherty Earth Observatory, calculated that the western shelf is in fact losing mass to the oceans. But whether this instability is the harbinger of an eventual collapse remains speculative.

One camp, the "dynamists," believes the western shelf has undergone rapid changes in the recent past and that it will continue to do so. Another camp, "the stabilists," believes the ice is unlikely to shear off or melt away, at least within the foreseeable future.

The Wordie shelf, floating on the coast of the Antarctic Peninsula, has been in the throes of breaking up for the past few decades, affording scientists the opportunity to study the complex interactions between the atmosphere, ocean, and ice. While experts debate whether Wordie's slow demise is attributable to regional warming or represents the initial stage of a continental meltdown, vital information has been gathered on the way the shelf becomes thinner, fractures, and disintegrates as the ocean and surrounding atmosphere warm.

They have also learned that this process is fraught with deception. When a general warming begins at either pole, the ice actually thickens. This is because the land and sea at the middle and tropical latitudes are heating up as well, producing greater amounts of water vapor, which is more potent at trapping heat than carbon dioxide. Some of this additional water vapor is carried by wind to the high latitudes, where it freezes and falls as snow. As long as the polar temperature remains below freezing, the snow will accumulate, feeding the ever-thickening glaciers. But if the warming continues until it reaches the melting point of ice, the process reverses itself and the oceans begin their inexorable rise, thus resolving a seeming paradox.

As mean global temperatures have advanced over the past century, atmospheric water vapor and other greenhouse gases have increased significantly. And as predicted, surface warming has accelerated at the poles. This is particularly true of the Arctic, where the trend is better documented and more pronounced. A thaw is already well under way in the Land of the Midnight Sun, according to a study conducted by Ola M. Johannessen, Einar Bjørgo, and Martin W. Miles of the Nansen Environmental and Remote Sensing Center in Bergen, Norway. Paralleling the warming of Toolik Lake, the arctic ice sheet has shrunk in both extent (−4.6%) and area (−5.8%) from 1978 to 1994. The dynamists see this as another of global warming's many signatures; the stabilists dismiss it as just one more nom de plume.

~

Few members of the public are aware of the menagerie of other greenhouse gases, numbering in the dozens. When trailed across the page, they appear to be some sort of enemy code: CH_4, N_2O, $O_3(T)$, $O_3(ST)$, F_{11}, F_{12}, F_{22}, CH_3CCl_3, F_{13}, CF_4, CH_2Cl_2, $CHCl_3$, CCl_4, C_2H_2, C_2H_6, SO_2. Taken together, these trace gases and others composed of carbon, fluorine, hydrogen, iodine, sulfur, and more, are thought capable of producing a global warming equivalent to that of carbon dioxide alone.

Third after water vapor and CO_2 in quantity comes methane (CH_4), 545 metric tons of which enter Earth's atmosphere each year, a figure that has been rising by about 1 percent annually over the last two decades. The concentration of CH_4 was about 0.7 parts per million four centuries ago. By 1988 it had increased to 1.7 ppm, largely because of the industrial revolution. If the current rate of increase is maintained, concentrations will reach 2.8 ppm in the next half century.

While atmospheric CO_2 is 200 times more plentiful than CH_4, molecule for molecule CH_4 is ten times more effective in absorbing and reradiating infrared energy back to Earth's surface. Odorless, colorless, and flammable, CH_4 is the major constituent of natural gas formed by the decomposition of ancient plant and animal matter. It is also produced by bacteria feeding on decaying organic material in swamps, peat bogs, and other natural wetlands.

As is true of CO_2, nature's production of CH_4 receives a significant boost from human activity. The growing of rice, the only grain used almost exclusively for food, results in the emission of some 12 to 14 percent of the world's annual CH_4 budget, about half of which is produced in India. Though rice can be grown on fairly dry uplands, the highest yields result from its cultivation under water, which in turn elevates the production of CH_4. Impoverished farmers mostly fertilize their fields with fresh organic matter, such as green manure and straw, further increasing CH_4 emissions. Yet for hundreds of millions who depend on rice for their survival, current production levels are inadequate and will have to be increased by nearly 50 percent over the next twenty years if widespread famine is to be averted.

Modern livestock-raising practices are also contributing to the elevation of CH_4 levels in the atmosphere, for the intestinal tracts of cattle and hogs serve as mini CH_4 factories whose product issues from both ends. Anywhere from 5 to 9 percent of what an average cow eats goes up in CH_4, compared to about 1.3 percent for the average pig. The foul-smelling slurry from vast feedlots and hundreds of hog pens, whose isolated occupants never once touch a cloven hoof to the ground until market day, is collected in giant, festering pools animated by rising bubbles of gas. It was once thought that the sacred cattle of India were equally to blame for CH_4 emissions, but it now appears that this calculation was too great by a third. Researchers forgot to take into consideration the fact that Indian cows eat and weigh considerably less than their North American cousins.

Minute by comparison but uncountable in their numbers are the nearly 2,000 species of termites, whose contribution to global warming has long been debated by scientists. Primarily wood eaters, termites convert the cellulose, a carbohydrate, into various sugars with the aid of CH_4-producing bacteria. These symbiotic organisms also thrive in the digestive tracts of millipedes, scarab beetles, and the ubiquitous cockroach, but not in all species. Tropical termites and cockroaches produce the largest amounts of CH_4, as do those living indoors in more temperate climates. Because an atmosphere rich in CO_2 stimulates plant production, a thicker biomass will almost certainly raise the output of CH_4 and provide more food resources for CH_4-producing bacteria. Moreover, researchers at the Catholic University of Nijmegen in the Netherlands think that millipedes, scarab beetles, and cockroaches, arthropods all, may contribute about the same amount of methane as do termites.

More troubling to some are the vast quantities of CH_4 locked in the frigid depths of the oceans. Among the concerned is James Kennert, an oceanographer at the University of California, Santa Barbara. Kennert has discovered evidence that temperatures in the Pacific rose as much as four to seven degrees Celsius in only a few decades during the last ice age.

The driving force, he conjectures, may have been a massive outburst of CH_4 gas. What triggered this event is open to conjecture, but the likely cause was a rise in sea surface temperatures that filtered down into the lower depths, destabilizing the environment.

This argument is bolstered by the discovery of rapid changes in the concentration of carbon-13 isotopes in marine sediments, which are consistent with a major gas release. Once the CH_4 rose to the surface, where it entered the atmosphere, a threshold was crossed and the climatic system ran amok until a new level of stability was finally achieved. Wallace Broecker thinks that this entire cycle may have been set in motion by increased surface warming due to elevations in the atmospheric water vapor. Moreover, the chance of it happening again remains significant so long as greenhouse gases continue to multiply, prodding the Tiger.

Such factors, together with the rest of the accumulating evidence, suggest to many scientists that Earth may well be teetering on the edge of another great climatic instability; that the time has come, as Carl Sagan concluded on that January night a decade ago, for us to exploit "the luxury of emulating Cassandra's listeners without sharing their fate."

17

SIGNS AND PORTENTS

~

May it not be an omen.

—an old saying

El Niño. The Boy Child. Enfant terrible. Named long ago by Spanish-speaking fishermen for the baby Jesus, it was once regarded as little more than a tepid current that visits the coasts of Peru and Ecuador at Christmas, temporarily reducing the rich fish stocks and giving the *pescadores* some welcome time away from the sea. Meteorologists were hardly more interested than the fishermen in what appeared on their charts as the Southern Oscillation.

All that changed forever during the winter of 1982–83, when the sea temperature off Peru rose by four degrees Celsius literally overnight. In the coming months, weather linked to El Niño would kill 2,000 people around the world and cause over $13 billion in damage from drought, flood, fire, and storm.

Nowhere was the suffering greater than in El Niño's host country. Peru's agricultural output fell by 8.5 percent, while production in its economically vital fishing industry was nearly halved. Meanwhile, inland and to the south, El Niño brought drought to the naturally dry Altiplano, the high plateau surrounding Lake Titicaca, which Peru shares with Bolivia. After their potato crops failed, desperate mothers attempted to give their children away to strangers visiting the local markets. More than a decade passed before Peru's recovery, bedeviled by terrorism and rampant inflation, was pronounced complete.

The story was much the same elsewhere, though the forms taken by the forces of destruction varied. Drought and fire struck southern Africa, India, the Philippines, Indonesia, Sri Lanka, Australia, Mexico, and Central America. Peru and Bolivia, already hit hard by drought, underwent rampant flooding, as did Cuba and Ecuador. In the United States, the most costly disaster in a year of weather-related catastrophes was a wave of Pacific storms that pummeled California's golden coast. The outburst triggered a series of deadly floods and mudslides, forcing over 17,000 families to register for assistance in centers operated by the Federal Emergency Management Agency.

When enhanced by computer graphics, El Niño resembles nothing so much as a doomsday virus of the type conjured up by Michael Crichton in *The Andromeda Strain*. A thick and angry red band representing superheated water spans the Pacific at the equator, reaching from the eastern shores of South America all the way across to Australia and Asia. Mottled oranges, blues, greens, yellows, and other vivid hues denote variations in wind speed, ocean currents, air temperatures, rainfall, and drought. The same colors appear on computerized maps of individual countries when the Child is on the loose. The winter will be warmer across the northern United States, the Gulf states cooler and wetter. California will suffer torrential rains and massive flooding; the Pacific salmon and other fisheries will be disrupted; the Atlantic Coast will be subjected to fewer, less violent hurricanes. (Equally dramatic but not as well known is La Niña, whose effects on the United States and elsewhere are usually the opposite of those of her infamous brother.)

El Niños occur every three to seven years and normally last twelve to eighteen months. Scientists measure the strength of these oscillations by how much warmer than normal the ocean waters become. Although they do not yet know whether the latest outbursts can be directly linked to the greenhouse effect, computer climate models indicate that the temperature rise in a warmer world should bring about an increase in such extreme weather.

At least two noted U. S. researchers do not believe that El Niño's recent rampages conform to a natural pattern. In 1996 Kevin E. Trenberth and Timothy J. Hoar of the National Center for Atmospheric Research pointed out that, starting in 1976, El Niños have popped up with a frequency unmatched in the last 113 years. In 1991 the Pacific spawned the longest-running El Niño on record; it lingered in equatorial waters through mid-1995.

When statistical tests are employed, both the frequency of El Niños and their duration over the last twenty years should occur only once every two millennia. According to Trenberth and Hoar, "This opens up the possibility that the El Niño changes may be partly caused by the observed increases in greenhouse gases." If they are right, the roughly equal pendulum swings between warm spells and cold ones in the Pacific are becoming unbalanced and will tilt in favor of more frequent, long-lasting Southern Oscillations.

Just such a development was postulated by Gerald A. Meehl and Warren M. Washington, colleagues of Trenberth and Hoar at NCAR. In a separate study published in *Nature* in July 1996, they conclude that sea surface temperatures in the tropical Pacific increased on average by several tenths of a degree Fahrenheit during the last two decades. "We cannot," they write, "definitively attribute the recent warming in the Pacific (an associated global warming of the 1980s and early 1990s) to increased levels of CO_2 in the atmosphere. Yet the model results presented here and elsewhere, as well as some observational results, point to the possibility that CO_2-induced climate change in the Pacific region could have this signature."

Others are more reserved in their judgments. Tim P. Barnett of Scripps takes issue with the test used by Trenberth. "I have problems, just statistically, with generating one million years of variation using 100 years' worth of data. It's an interesting analysis, but it doesn't convince me that this thing is as rare as what he's saying."

Close on the heels of this exchange came the next El Niño, the most powerful and widespread of the twentieth century. This time meteorologists were not about to be caught off guard and cautioned the world to prepare itself for a protracted and potentially devastating siege. To the misfortune of many, the long-term forecast was dead on the money.

~

In the Galápagos Islands, the cool misty season known as *garua* never materialized in June of 1997. Instead, the climate grew hot and sticky, giving Darwin's normally arid paradise the feel of the Amazon. In November came the rains, which in a typical year don't begin until February. Driven by trade winds seemingly gone mad and fueled by energy drawn from ocean waters warmer than average (by three degrees Celsius), they pummeled the score of volcanic islands for months on end, as if God had tired of Creation and wished to begin anew.

The number of flightless cormorants had fallen by 45 percent during the El Niño of 1982–83. The species, unique to the Galápagos, was now under serious threat again. So, too, were the rare Galápagos penguins, whose numbers had dropped by nearly 80 percent and were just recovering fifteen years later. As before, waved albatrosses failed to breed, while, in the sea, green algae, the main food of the marine iguanas, were replaced by an inedible red species, triggering the lizards' starvation. In just two days in mid-December, rainfall on the island of Española totaled 204 millimeters and was rapidly gaining on the 2,770 millimeters of precipitation—six times normal—that had scoured the Galápagos in the early eighties.

Almost overnight, these barren volcano tops, some of which are still in their birth throes, became luxuriant with vegetation. Even the most arid islands were covered with a verdant carpet, an idyllic but troubling sight to resident biologists who believe that such luxuriance poses the most serious threat of all to the biodiversity of the Galápagos. Conservationists have long been doing battle with nonnative species—rats, cats, goats, fire ants, and biting flies—all of which thrive in surroundings rich in plant life. On the positive side, the native tortoises benefit from an abundance of vegetation, as do Darwin's well-studied finches, which breed as many as three times a year during an explosion of the seed-bearing flora.

Kicking and screaming, El Niño began flailing its way across South America in August of 1997. While blizzards enshrouded parts of the Peruvian Andes in record snowfall, stranding travelers and freezing them to death, Rio de Janeiro recorded its hottest winter day—108ºF—in seventy-five years. Chile's rat population, linked to a killer hantavirus outbreak, exploded in the wake of unusual plant growth, especially the flowering of the bamboo. By January of 1998, Peru's snows had given way to torrential rains, dropping as much as thirteen liters per square meter in a fourteen-hour period. Within weeks fifty-nine bridges collapsed and 530 miles of highways were destroyed. The government, fearful of further depleting the dwindling stocks of anchovies, imposed a three-month ban on commercial fishing. Other normally abundant species all but disappeared, having followed the plankton into cooler waters.

In Ecuador the number of cholera cases reached 3,084 in mid-February of 1998, surpassing that for all of 1997, while floods and landslides killed 108 and drove another 28,000 from their homes. Some 37,000 acres of banana trees were inundated, ruining one of the country's key exports. Countless ponds became the breeding grounds for diseases ranging from sores on the limbs of children to mosquito-borne malaria, encephalitis, and dengue fever.

Half a world away, on the Horn of Africa, the normally docile Juba and Shabeele Rivers merged after three weeks of El Niño–spawned rains to create an inland ocean unlike anything Somalia's pastoral population had ever seen. Two thousand people died, and tens of thousands of animals were swept away. In the deadly competition that followed, marooned Somalis were attacked by hyenas as man and animal fought desperately to claim what little was left of the high ground. The waters had barely receded when the nation's 250,000 homeless were struck by the largest-ever outbreak of Rift Valley Fever, incubated in the residual "virus soup." As in the aftermath of the hemorrhagic plague that struck ancient Athens during the Peloponnesian War, wary canines and vultures left the corpses untouched.

Neighboring Kenya endured similar travail. The worst flooding in the country's history left at least eighty-six dead and devastated the nation's roads and bridges. Sheep farmers were hard hit by an outbreak of bluetongue, a disease last seen in 1905. Losses ranged from 10 to 33 percent of the sheep population and were compounded by a drastic falloff in coffee and grain production. Among the little good news was a report that normally brackish Lake Nakuru in the Rift Valley, freshened by rains, had attracted 1.5 million flamingos.

So powerful had El Niño become by late 1997 that scientists had begun to wonder whether it was actually slowing Earth's momentum. During non–El Niño years, winds in the tropics blow from east to west, whereas winds over the rest of the globe travel from west to east. When combined, they give the atmosphere a net eastward momentum. But El Niño slows down the tropical easterlies and speeds up the westerlies, boosting the atmosphere's angular momentum, a property comparable to the impetus of a spinning tire. Over a typical year, the day shortens and lengthens by roughly one millisecond. During the 1997–98 El Niño, the day grew longer by four-tenths of a millisecond, a minute but profound change.

Few caught in El Niño's wake were in a position to argue the finer points of physics. Beginning in September of 1997, a milky twilight, brought on by smoke and haze, settled over the Malaysian capital of Kuala Lumpur each day around noon, transforming the tall buildings into ghostly shadows. People scurried along the streets, their noses and mouths covered by disposable masks, as if the world had become a giant operating theater and every citizen a surgeon.

In the state of Sarawak in Malaysian Borneo, the level of pollution rose above that of London during the Great Smog of 1952 and was equated with smoking eighty to one hundred cigarettes a day. In the midst of rising death rates, schools and airports were closed, but despite such precautions the number of fatalities mounted. Ship collisions in the Strait of Malacca killed 29 people, while haze from Indonesian fires was suspected in the crash of an airliner that claimed 234 lives. Traffic accidents blamed on poor visibility killed hundreds more. Meanwhile, tens of thousands suffering from asthma, vomiting, diarrhea, and fever lined up at hospitals and clinics, whose medical personnel were overwhelmed. Tourists canceled their reservations, cutting off a critical source of foreign exchange; crops short of rain and sunlight withered; doctors at the World Health Organization pondered future cancer rates. In outlying villages tigers attacked peasants, who in turn vented their wrath on starving and displaced orangutans, clubbing and spearing the shaggy "man of the forest" to death.

By October fires intentionally set by humans and aggravated by drought attributable to El Niño had consumed an estimated 4.2 million acres of rain forest across Southeast Asia. Scores of agricultural companies with high-placed government connections started their illegal burning in June, at the onset of the dry season. By August almost every smoke detector for hundreds of miles around had to be disconnected to maintain sanity. With El Niño on the prowl, the monsoons of autumn, the only sure way of dousing the flames, failed to materialize. This came on top of increased emissions from factories and vehicles that had already caused a rapid deterioration in the region's air quality with the onset of modernization.

Australia's turn came next. By November hundreds of brush fires were out of control in the states of New South Wales and Western Australia, consuming more than 85,000 acres and killing 10,000 sheep. Some 7,500 firefighters, backed by military units, were thrown into the battle that included the evacuation of 20,000 people. A wall of flames up to forty meters high swept across the suburbs of Sydney. By the time it was over, insurers considered themselves fortunate that "only 200" homes were lost.

With sea surface temperatures warmer off Australia's eastern coast than during the 1982–83 El Niño, coral bleaching along the Great Barrier Reef renewed scientific concerns for the "rain forest of the sea." The bleaching was first observed on numerous inshore reefs, after water temperatures reached a critical 30°C, two degrees higher than normal. As the ocean warmed, the bleaching moved north and east, reaching New Caledonia, some 800 miles from the Australian coast, in December of 1997. Scientists from the National Oceanic and Atmospheric Administration (NOAA) also confirmed the return of coral bleaching to the Galápagos, as well as the coasts of Ecuador, the Florida Keys, Baja California, Panama, the Yucatán, the Caymans, and the Netherlands Antilles, a discordant note on which to ring in 1998, the "Year of the Ocean."

Whether a mutant created in part by humans or simply one of nature's bad jokes, El Niño was far from finished. As if mocking those who had named it, the Child swept through

New Mexico on Christmas Day, 1997, subjecting the state's ranches to a barrage of icy wind, snow, and rain. Hardest hit was the Pecos Valley in the southeast. On a single ranch called the B23 Cattle Company, the carcasses of 3,000 calves lay in clumps that at first glance appeared to be snow-covered hills. Then the melting began, moving the ranch's owner to observe: "It looks like a nuclear disaster has swept the area. We're broke."

Three weeks later the most savage ice storm on record ripped into upper New England and northern New York before crossing into Canada, where it wreaked havoc on five eastern provinces. Uprooted trees and severed branches littered forests and city parks alike, after which came night and the paralyzing darkness. Across the region utility poles and electric towers crumpled to the ground, as if flattened by some gigantic beast. Days passed, and still the power did not flow, forcing tens of thousands to take refuge in public shelters or huddle with neighbors who had a ready supply of firewood.

The death toll climbed as many, including the elderly, refused to leave their freezing houses and apartments; others lost their lives in fires they themselves had started, or succumbed to carbon monoxide fumes from poorly vented heaters. Supplies of candles, kerosene, camping lanterns, wicks, and batteries quickly disappeared from the few stores that remained open. Signs along the major highways warned motorists that gas stations were closed for a lack of electricity to run the pumps.

It was a Robert Frost nightmare of birches split down the middle by their own glazed weight, of temperatures stubbornly refusing to rise, of cranky New Englanders growing ever more cranky waiting for the power to come back on, of unmilked cows dying of infection, of a world frozen in time. "This isn't normal," remarked one weary septuagenarian after spending six days in a shelter playing cards and worrying about her deserted home. "It's like the end of the world."

Only months before, fishermen in the Pacific Northwest had noticed that something strange was happening. A quarter of the king crabs being held in seawater tanks died in one four-day period, compared with a normal loss of 1 percent. A temperature test revealed that the water was warmer than normal by four to six degrees Fahrenheit. Albacore, a major source of canned tuna, appeared along the Alaskan coast much earlier and in far greater numbers than usual, causing the fishing industry to launch 200 vessels, six times as many as the previous year. In California's Sacramento River, the fall run of Chinook salmon began in September, two months ahead of schedule, and was attributed to Pacific waters several degrees above average. Near San Francisco, a twelve-and-a-half-pound jumbo squid, *Docidicus gigas,* was pulled live from the normally chill ocean. All along the California coast sportsmen not only were enjoying unprecedented success but were hauling in trophy fish of species rarely if ever seen so far north.

Beginning in February 1998, El Niño gave rise to a series of storms that dumped six feet of rain on a coastal region that normally receives about fourteen inches. California's flooding and mudslides became a regular feature of the evening news as people waited for the storm that would topple the next million-dollar house from its exalted perch. So much the better for ratings if it fell onto another below it before both were sacrificed to an angry sea in a cascade of bricks, splintered timbers, and broken glass.

While the much-envied and despised California lifestyle received most of the media's attention, the chaos was mainly rural. Eleven months out of twelve, the battle for farmers living in the Central Valley is to irrigate their arid soil. Within a week of El Niño's arrival, they were wading through fields that had become a mixture of mudflats and shallow lakes. Strawberries, ripe and exploding with moisture, split on the vine, requiring that they be ground into pulp. Artichokes, lettuce, spinach, and cauliflower were either washed out or drowned, sending prices through the roof. Pickers who normally labored under a hot Sun worked through the darkness and pounding rain in a futile attempt to sandbag the fields. Many of them would soon draw their last paychecks before facing long stretches of joblessness.

A parallel cycle of destruction was unfolding across the continent. Tornado Alley, long associated with the Plains states, temporarily shifted to the south and east. In February the worst tornado outbreak in Florida's history claimed forty lives and caused $67 million in damage. Winds reached an estimated 260 miles per hour and were duplicated in twisters that hit Georgia, South Carolina, and Tennessee weeks later. By May the death toll had risen to 112, yet another example of El Niño's deadly caprice.

~

The time had come for climatologist Hugh Ellsaesser of Lawrence Livermore National Laboratory to make out a check for one hundred dollars, payable to James Hansen. Six months earlier, during the summer of 1990, Hansen had offered to bet all comers that one of the first three years of the new decade would be the warmest ever recorded. As the only researcher to accept Hansen's wager, Ellsaesser admitted to defeat in early January of 1991, after three different measures of global temperatures all proclaimed 1990 the warmest year in the history of record keeping. It had easily surpassed 1988, the previous record holder, when Hansen had warned the Senate Committee on Energy and Natural Resources that global warming was at hand.

"It doesn't change my opinion of what's going on," a game Ellsaesser maintained, and many climatologists agreed with him that a short and still-modest warming cannot be conclusively linked to greenhouse gases. Hansen himself concurred that one year alone, even a year that sets a new record, is unimportant because global temperatures fluctuate.

Still, "it was a good bet," if not 100 percent certain. Hansen reasoned that it would be very difficult for Earth's climate to cool significantly for any extended period as long as the atmosphere is subjected to increasing amounts of carbon dioxide. A single year of unusual heat could be a fluke, but the underlying trend was in his favor. What made the temperature milestone even more notable is the fact that it had occurred without a contribution from El Niño, whose surge of warm water had helped boost 1987 into the record book, only to be surpassed by 1988.

Hansen also knew that six of the seven warmest years in more than a century had occurred since 1980. With the same data available to other climatologists, it is hardly surprising that nobody but Ellsaesser had taken Hansen's bet. Common sense, if nothing else, dictated that the upward movement of

James E. Hansen
(Photograph by Gabriela Porter)

temperatures would continue, so why subject oneself to the double indignity of throwing good money away while enduring the gibes of one's peers? Hansen could take heart from the fact that his colleagues' refusal to wager was a tacit admission that he seemed to know what he was talking about.

The pulse of warming quickened as the 1990s wore on. In July of 1995, Chicago was struck by a deadly heat wave. Apartment dwellers whose windows had been sealed tight against the winter elements were helpless as their brick residences were transformed into human kilns. Evening temperatures hovered near their daytime highs of 100°F, and at least 522 people died within days. Because nothing approaching this heat wave had ever happened before, the best that benumbed public officials could do was to provide the suffering with fans.

In the summer of 1996, the southwestern United States was in the grip of the worst drought since the Great Depression. Falcon Lake, which covers some 87,000 acres along the Rio Grande and the U.S.-Mexican border, was created in the 1950s to improve flood control and irrigation. As the water level slowly dropped by almost fifty feet, secrets of its long-submerged past were exposed bit by bit. On the Texas side of the lake, the drowned border towns of Zapata and Lopeño reappeared. On the Mexican side, near the town of Benevides, the stone crosses of a flooded cemetery rose like eerie sentinels in a Stephen King novel. The last time anyone had seen these graves, segregation was the law of the land and the Dodgers were playing baseball in Brooklyn.

Conditions were the same across southern California, southern Nevada, all of Mexico, Arizona, and Texas, and much of Oklahoma, Colorado, and Utah. Some parts of the region had been afflicted for up to five years and other areas for as few as several months. From August of 1995 to June of 1996, great swaths of the Southwest and the southern Plains recorded little more than a trace of rain or snow. The failure of the winter wheat crop led to a shortfall in the supply of cattle feed, forcing ranchers to liquidate their herds in record numbers. Oklahoma's commissioner of agriculture predicted that 5,000 to 10,000 of the state's 70,000 farm families would go bankrupt within a year; from Kansas to south Texas additional thousands of farmers were on the verge of financial ruin and joined in the panic to sell their livestock before there was nothing left to sell. The glut was so great that auctions beginning at midmorning continued through the night, the bidding per head averaging a mere twelve seconds.

As it always does when water in the region becomes scarce, talk turned to the Dust Bowl. In the Oklahoma Panhandle, which was enduring the second-driest period since the beginning of record keeping in 1895, seventy-one-year-old farmer Lewis Mayer was about to plow under his failed wheat crop. He crumbled a lump of warm soil between his fingers while standing in 105° May heat and recalled that the last time such temperatures had been recorded, he was "a ten-year-old boy in 1935. . . . [this soil] just looks similar. Very similar."

Later in July, it was massive flooding in eastern Europe that claimed more than a hundred lives. In August Typhoon Winnie ravaged the Chinese coast near Shanghai, forcing more than a million residents to flee. In October the worst wildfires in thirty years devoured 6 million acres in the southwestern United States.

If global warming was the cause, many scientists are fearful that such record-breaking disasters will become more frequent and more extreme in the future. However, they are not able to say that any single weather event is the result of global warming. Looking back to 1990, even a confident James Hansen might have been a bit surprised to learn just how safe his wager had been. If he had not collected that year, he need only have waited until 1991, when another record was set, followed by still another in 1995. In fact, the period from 1991 through 1995 was the warmest half decade ever recorded, surpassing 1986 through 1990. This notwithstanding the June 1991 eruption of the Philippines's Mount Pinatubo, whose microscopic debris cooled the globe for three years before drifting back to Earth. In January of 1998, climatologists announced that nine of the warmest years on record had occurred in the last eleven years.

Moreover, 1997 was recently proclaimed the warmest year to date by NOAA. Like much else, this dubious distinction is at least partially attributable to the manic El Niño. Yet even if El Niño had not been a factor, 1997 would still have been a very warm year, extending a trend that has made the 1990s Earth's warmest decade since thermometers were first used to measure temperatures in the mid–nineteenth century.

The planet's average surface temperature is now about 62°F and is rising by about ten-hundredths of a degree per decade, which breaks down to slight hundredths of a degree per year. In an interview with the *New York Times* science reporter William K. Stevens, Hansen noted that such small increments are "not really that significant" in helping to judge whether human activity is warming the atmosphere. A significant rise globally, such as one-tenth of a degree Celsius (nearly 0.2°F), "would make it clearer. I suspect we will see that in the next couple of years." Stevens did not ask the NASA scientist if he was still taking bets.

18

SCENARIOS

~

They set an ambush for their own lives.

—Proverbs

It was by coincidence that only a week after James Hansen delivered his dramatic testimony before a Senate committee in
Washington, delegates from nearly fifty nations assembled in
Toronto in May of 1988. Their agenda: to address ways of
bringing Earth's temperature back under control. It was the
first International Conference on the Changing Atmosphere.

The talks went so well that the same delegates met in
Geneva that fall to form what would become the world's leading authority on global warming, the Intergovernmental Panel
on Climate Change (IPCC). Some of the panelists came from
the countries most responsible for the release of greenhouse
gases: the United States, the U.S.S.R., Great Britain, West Germany, and Brazil. Others came from states that contribute
very little to the problem but have a great deal to lose, including Malta and the low-lying Maldive Islands. Together they
created a scenario of what the future may hold if nothing is
done to curb carbon dioxide emissions. Now an organization
sponsored by the United Nations, the IPCC is made up of
2,500 scientists from around the world, and it is their method-
ical and detailed analysis of global warming that dominates
the climate debate.

The IPCC's *Climate Change 1994* is the most authoritative

report ever produced on the subject. It was composed by 78 "lead" authors, among them the world's foremost climate scientists, with input from 400 more. Another 500 scientists and policy experts scrutinized the early drafts. Finally, government delegates from ninety-six countries, along with a small army of technical advisers and representatives from industry and environmental groups, met in Madrid in 1995 to discuss the draft report, propose further changes, and approve, line by line, the official summary of its conclusions.

~

Imagine, for a moment, that the clock governing civilization has been advanced to a summer's day a century into the future. Carbon dioxide levels, which stood at 370 parts per million on the Keeling curve in the year 2000, are nearing 950 ppm. On a computerized weather map of the United States, the day's low temperature of 60°F is predicted for Point Barrow, Alaska, the high of 131° for Palm Springs, California. The northern two-thirds of the country will suffer through yet another in a long string of days in the high 90s and 100s, while Little Rock, Atlanta, Birmingham, Houston, and Albuquerque will register highs of 105° and above. Dallas, San Antonio, Tucson, and Phoenix will top out at 125°. Miami would likely have done so as well but for the fact that the city, indeed the whole of South Florida below West Palm Beach, is no more because of a dramatic rise in the sea level. What is left of the state can expect gale-driven rains, softball-size hail, and increased beach erosion. A health advisory has been issued by the Centers for Disease Control; the mosquito populations continue to multiply, spreading illnesses against which vaccines are ineffective.

Along the coasts of Georgia and South Carolina, where the barrier islands have long since been devoured by the Atlantic, ten- to twelve-foot swells are expected to cause severe flooding, leading forecasters to recommend the evacuation of low-lying areas. Tropical Storm Jeremiah has been upgraded to Hurricane Jeremiah and will likely make landfall in the vicinity of Wilmington, North Carolina, in another forty-eight hours, packing sustained winds of 245 miles per hour.

Southern California is told to brace for more mudslides due to continued torrential rains, while in the Dakotas brush fires are rampant, posing a threat to the great herds of buffalo that have reclaimed their native range, covering the land like a moving robe. The central Plains have experienced less than 0.04 inches of rainfall in the last six months. Food prices are projected to rise sharply for the fifth year in a row, significantly impacting consumers' pocketbooks and causing panic on the trading floors of New York and Chicago.

Those contemplating a weekend getaway are urged to consider New England. Pleasant weather is expected, with clear skies and high temperatures only in the mid to upper 90s. Unless gambling is one's passion, Nevada is to be avoided at all costs. Las Vegas will check in at 118°, Reno at 114°. A Weather Service Advisory recommends that residents of the state receive minimal or no exposure to the Sun.

The global outlook is just as dire. All four hemispheres, North and South, East and West, are stewing in their own juices. Mudslides and 100° are predicted for Hong Kong. Drought-stricken Jerusalem will reach 117°, with equally parched Nairobi, Athens, Beijing, and Istanbul not far behind. Those who love Paris when it sizzles are in luck; 100° and flooding are the order of the day.

~

Few who are members of the IPCC would embrace such an apocalyptic scenario, although many environmental organizations believe that such developments are within the realm of the possible. There is no question, however, that many of the fundamental ingredients of such a global catastrophe are contained in the climate change reports issued by the IPCC, documentation that invites closer scrutiny.

Satellite data first became available in 1979 and currently provide the only truly global temperature measurements. Critics of the greenhouse effect have been quick to point out that these readings actually declined slightly over the past two decades. Meanwhile, proponents of the global warming hypothesis have mostly relied on data gathered by weather stations at Earth's surface, according to which temperatures have steadily risen, and continue to do so. This disparity between a cooling atmosphere and a warming planet seems to have been resolved by Frank J. Wentz and Matthias Schnabel, atmospheric physicists employed by Remote Sensing Systems, a private research firm located in Santa Rosa, California.

In a 1998 article in *Nature*, Wentz and Schnabel explain

how, in their reanalysis of the satellite record, they uncovered a significant distortion. Scientists had failed to take note of the inevitable decay, or lowering, of the satellites' orbits as they encounter atmospheric resistance, the effect of which is to lower the temperature readings taken by the onboard instruments. Once this error is corrected, the satellite record reveals a warming of about 0.07 degree Celsius per decade, bringing it in closer agreement with surface temperatures. When asked to comment on the importance of this discovery, an elated James Hansen proclaimed that it could signal nothing less than "a sea change in the global warming debate."

As a rule, warmer climates have more uniform temperatures than do cooler ones. The mercury doesn't rise and fall as much at the equator as it does at the higher latitudes. Thus if greenhouse gases are warming the entire globe, temperature ranges should be narrowing, literally reducing the age-old difference between night and day. According to IPCC scientists, this is precisely what is happening. Climatologists studying several decades of weather records for China, the United States, and the former Soviet Union have found that across the Northern Hemisphere temperatures currently do not swing as widely as they once did in a day-to-day, week-to-week, or month-to-month time frame. The same records, which date back twice as far in the United States as elsewhere, also indicate that precipitation now falls more often in extreme bursts of two or more inches in a day than it did a century ago, which is in line with computer models. Whether this is true outside North America remains unclear.

It was Darwin's contemporary Sir Charles Lyell who first argued that the survival of any species is very much dependent on the stability of its environment. As conditions change over time, a species may respond by shifting habitats and thrive, or it may perish if the pressures to which it is subjected are too great. Though he recoiled at the thought of evolution, Lyell could not have been closer to the truth.

A great many plant and animal species vanished as the global temperature rose by a mere degree or so after the retreat of the vast ice sheets. In North America alone mastodons, saber-toothed cats, giant armadillos and ground sloths, native horses and camels—indeed, almost three quarters of the continent's large mammals—became extinct. Spruce forests shifted northward and grew denser as hardwoods and open grasslands took their place.

Recent studies have shown that trees are extremely sensitive to temperature changes. Even a warming of one degree Celsius could cause the southern borders of forests throughout the country to move northward. States such as Georgia and Mississippi could see their woods completely disappear, only to be replaced by heat-tolerant grasses. In the northern states of Michigan and Minnesota, trees such as the balsam fir and birch might die out, leaving behind scattered oaks and a few other hardwoods.

If the climate changes slowly enough, warmer temperatures will enable trees to move north into areas that were previously too frigid, at about the same rate, one would hope, as the southern areas become too hot and dry for their survival. It is estimated that if the planet warms by two degrees Celsius (3.6°F) in the next century, species will have to migrate at a rate of two miles per year. Trees whose seeds are carried by birds may be able to do so, although the prospect seems dubious at best. Others, including nut-bearing species such as hickory, hazel, and oak, are unlikely to spread more than a few hundred feet annually. And this assumes that arboreal migration routes are free of impediments such as highways, construction, and other modifications to the environment. On balance, losses are likely to occur more rapidly than expansions, a process that would accelerate if the climate becomes drier.

Many animals will vanish as well, beginning with several species of birds whose nesting ranges become ever more constricted. A primary candidate is Kirtland's warbler, a small yellow-and-gray songbird that nests in the sandy, well-drained soil beneath the low branches of young jack pines in north central Michigan. These trees have an extremely low tolerance for heat and will likely disappear as quaking aspen and oak encroach along the southern margin of their range. The warblers cannot shift their nesting sites because the compact soil beneath more northerly jack pines causes their nests to flood during heavy rains.

Least threatened will be the plants and animals living at

high elevations. Not only have the mountains escaped much of the development and fragmentation sustained by other regions, a relatively short migration up a slope lands an organism in a substantially cooler microclimate. For example, a climb of only 1,500 feet offsets an increase in temperature of five degrees Fahrenheit. On the other hand, a mountain's topography imposes physical limitations: The higher a species goes, the less the available space. As these cool islands shrink under a rising sea of warm air, so too will the number and diversity of the plants and animals inhabiting them. Species that were once free to migrate across continents will become isolated and inbred, their lack of genetic variety becoming as deadly as the rising temperatures that forced these changes to begin with. Complex interactions between predator and prey, host and parasite will likely spin out of control, triggering a cascading effect that will severely damage, if not destroy, entire ecosystems.

This threat comes at a time when biologists believe that 30,000 species are already becoming extinct each year, the equivalent of three species lost every hour—forever.

The one life-form certain to flourish in a warmer environment is the insect population. Illnesses spread to humans from species such as mosquitoes and ticks are called vector-borne diseases. Among the most threatening of these are malaria and dengue fever, both of which followed in the wake of El Niño. The World Health Organization places the number of people currently infected with malaria at between 300 and 500 million, of which 2 million, the majority of whom are children, die each year. At the moment, the types of mosquitoes that carry the disease can survive only in a warm environment, so that cooler temperatures act as a barrier against their spread. This is likely to change dramatically with a rise in temperatures. An additional 46 to 60 percent of the global population could be living within the malaria transmission zone by the latter half of the twenty-first century. Because populations in much of the Northern Hemisphere will lack natural immunity, the mortality rates will be high, at least to begin with.

Dengue, or "bonebreak," fever is spread by the mosquito as well. Characterized by high fever, chills, severe headache, and acute pain in the joints, it too has been confined to the tropics and subtropics for the most part, although cases have been reported as far north as the southern United States. There is no specific treatment other than good nursing care, the absence of which can kill as many as 15 percent of those infected. A temperature increase of three to four degrees Celsius could double the transmission rate of the virus.

~

In parts of the western United States, the most widely discussed aspect of global warming is its potential effect on the freshwater supply. If the climate becomes hotter and drier, natural runoff from rain and melting snows could decline by as much as 15 to 20 percent in places like the Colorado River basin. Yet the waters of the Colorado, like those of several other rivers, are already fully allocated.

The problem is compounded by the fact that a significant fraction of western water is used by irrigation farmers who pay only a few dollars per acre-foot for water delivered by federal projects, less than the delivery cost itself. At the same time, some municipalities are paying hundreds of dollars per acre-foot for the identical resource, an imbalance that will have to be addressed as temperatures rise. Will farmers, whose crops are needed to help feed a global population of more than 6 billion, be forced out of business? If not, who will pay the additional hundreds of millions of dollars annually to keep the tractors, the sprinklers, and the cooling systems for the power plants going?

Entwined in this Gordian knot is the question of hydropower. As a general rule, a 1 percent decrease in runoff results in a greater than 1 percent decrease in power production. Not only does less water pass through the turbines, lower reservoir levels reduce the water pressure, hindering power generation. In the Colorado's lower basin, for example, a 10 percent decrease in runoff reduces electrical production by 36 percent. Other studies conducted on the Great Lakes, the Tennessee Valley, the Missouri River, and the Atlanta area suggest that all will undergo a decline in output as the atmosphere warms. Thus hydropower, which is one of the two major sources of electricity that do not emit carbon dioxide, will almost certainly decrease in the coming years, spurring the demand for fossil fuels. Meanwhile, nuclear power, the other nonpolluting source, has little support from a public that retains vivid memories of Three Mile Island and Chernobyl, although that could change with the generations.

The eastern United States, where the water supply is sufficient, at least for the time being, confronts a different dilemma. As precipitation declines, estuaries, rivers, and wells are subject to increased salinity, a problem accentuated by a rise in the sea level. Florida has already spent billions constructing an elaborate network of barriers to protect its freshwater sources inland from the restless Atlantic. If these barriers were to fail, salt water would flow up the canals and gradually permeate the ground around them. In time, there would be no fresh drinking water on the coast. The water supplies of New York City and Philadelphia are also in jeopardy. As regional sources undergo contamination, delivery costs will increase, together with the necessity of expanding water storage capacity as a hedge against the unpredictable. Another score of the world's coastal cities are similarly threatened, including Amsterdam, Rotterdam, Liverpool, Istanbul, Venice, Barcelona, Gothenburg, St. Petersburg (Russia), and Archangel.

As the heat spreads, the problem of water availability will be complicated by the need to irrigate. In the Corn Belt, which stretches hundreds of miles across the Midwest, land that for a century and a half has been cultivated by relying solely on rainfall will probably have to be watered from June through August. Increased evaporation will also call for more irrigation in areas already being artificially supplied. Prices will rise accordingly, marking the end of inexpensive produce and putting pressure on the livestock industry, dependent as it is on large quantities of cheap feed grain. Sooner or later, the great underground aquifers will go dry, like wrung-out sponges, but perhaps not before they are permanently tainted by agricultural chemicals and the animal waste generated by giant feedlots and hog farms.

In all the various climate change scenarios advanced by the IPCC, world cereal production will decline, but no one knows by exactly how much. The one possibility of offsetting crop reductions again involves the costly operation of irrigation equipment. Climate change was also found to increase the disparities between the developed and developing nations, with the latter getting the short end of the stick. In the Northern Hemisphere, certain aspects of global warming will likely enhance agriculture. The longer time between the last frost of spring and the first frost of autumn will extend the growing season, of particular benefit to parts of Europe, the former Soviet Union, and Canada. Carbon dioxide absorbed during photosynthesis will act as a fertilizer and spur the growth of certain plants. On the other hand, some cereal crops need cold winters to initiate flowering. Warmer temperatures could stunt plant growth and lead to reduced harvests. In tropical and subtropical regions a further rise in temperatures will almost certainly reduce crop yields. In India, Bangladesh, and other parts of Asia, higher sea levels may claim low-lying farmland and cause higher salt concentrations in the coastal groundwater.

~

At night, when he is sound asleep, Teunaia Abeta can hear the lapping of the sea in his dreams. When he awakens in the morning, the lapping continues, as though he is living a dream within a dream.

A citizen of the Pacific island nation of Kiribati, he sits on the raised platform of his home dressed only in his lava-lava, a colorful skirtlike garment, smoking a hand-rolled cigarette. In all his seventy-three years, he tells reporter Nicholas Kristof of the New York Times, he has never seen anything like it. On a sunny day in January 1997, a tide came rolling in from the turquoise lagoon and did not stop on reaching the shore. It lapped higher and higher until it swallowed Abeta's thatched-roof home and dozens of others belonging to his neighbors. A month later, a second tidal surge added to the destruction of the first. Again, there was no wind, no rain, no warning of any kind. The ocean just seemed to swell, like so much liquid yeast.

The rising of the seas is linked to global warming in two ways: first, by the melting of the polar ice, and second, by causing the water itself to expand, for warm-water takes up greater space than cold. If this process continues well into the twenty-first century, the swelling seas will have reached a height unknown to past civilizations.

Kiribati, like many of its sister countries in the Pacific, is a collection of coral formations that resemble sandbars more than sovereign nations. Each island rises only a few feet above sea level and is roughly a hundred yards across, about the same as a football field from end zone to end zone. Many homes enjoy views of both coasts, and when the tide is out the islands briefly triple in size.

They are also the most likely to be engulfed by the sea, as are the central Pacific's Marshall Islands and Tuvalu, together with the Maldives, located on the Indian Ocean. And while they will not completely vanish beneath the waves, the island nations of Tonga, Nauru, Niue, Palau, and the Federated States of Micronesia are liable to lose much of their scant territory. The irony is all the greater because their citizens' only resource is water, and they did nothing to foster the conditions that may seal their fate. Notes a saddened and bitter Teburoro Tito, president of Kiribati, "It's like little ants making a home on a leaf floating on a pond, and the elephants go to drink and roughhouse in the water. The problem isn't the ants' behavior. It's a problem of how to convince the elephants to be more gentle."

The elephants themselves may not be that far removed from danger in a world where additional inches of water can obliterate millions of acres and two or three feet can claim entire countries. The IPCC projects a sea rise in the neighborhood of twenty inches by 2100, although some of its members believe that the real figure could be much higher. A rise of three feet in sea level would force the evacuation of an estimated 70 million Chinese and 30 million or so Bangladeshis, a displacement of epic proportions. Dutch tenacity notwithstanding, there is no guarantee that Holland's dikes will hold. In the winter of 1953, the raging North Sea breached the defenses in eighty-nine places along the central delta, drowning nearly 2,000 people and tens of thousands of cattle.

The entire East Coast of the United States, large stretches of which are already in jeopardy, would be subject to major erosion and deadly tidal surges, as would the Gulf Coast states of Florida, Alabama, Mississippi, Louisiana, and Texas. The low-lying cities of Galveston, Mobile, St. Petersburg, Sarasota, and a score of others are directly in harm's way, threatening the major insurance companies with bankruptcy.

The coup de grâce might be delivered by a series of monster hurricanes fed by the same energy that sustained the latest incarnation of El Niño. The Hugos and the Gilberts, the most powerful such storms on record, may become more powerful still. Consider what Hurricane Andrew, with an average wind speed of only 145 miles per hour, did to the community of Homestead, Florida, in August of 1992. The city looked like a nuclear Dresden, and the final bill, if there is such a thing, exceeded $16 billion.

Picture hurricanes packing sustained winds of 250 to 300 miles per hour, twice those of Andrew, taking dead aim at the Florida Keys, Miami, Jacksonville, Savannah, Charleston, or Myrtle Beach. No ordinary names would do them justice; only those of the most vengeful gods would suffice.

Whether the sea level rises by one foot or three, coastal marshes and swamps, commonly known as wetlands, are particularly vulnerable. Thousands of square miles would be lost to the surging waters, while property owners and municipalities scramble to create seawalls, jetties, dikes, and other structures in an attempt to fend off the inevitable. The Environmental Protection Agency estimates that a two-foot rise in sea level will eliminate between 17 and 43 percent of U.S. wetlands, with more than half the loss taking place in Louisiana alone. While new wetlands will form inland as previously dry areas are flooded by higher water, the size of the new marshes will be much smaller than those that are lost. The impact on wildlife, particularly birds and fish, will be major, bringing some species to the verge of extinction and pushing others over the edge. Many public recreation areas, including state and national parks, will simply vanish; the rest will be subject to periodic flooding and siltation, destroying their aesthetic appeal. Coastal engineers estimate that a one-foot rise in sea level will cause beaches to erode one-half to one foot from Maine to Maryland, two feet along the Carolinas, one to ten feet along the Florida coast, and two to four feet along the California coast. Even a one-foot rise would threaten homes in these areas, placing the government under intense pressure to save the American dream at any cost.

Coastal marshes are the primary nursing grounds for crab, shrimp, menhaden, and other species, with most reproduction taking place within fifty to sixty feet of open water. As the sea level rises, the loss of marsh accelerates until the protected ecosystem merges with the ocean and reproduction ceases. Oysters, which thrive in estuaries, will be subject to the triple whammy of higher water temperatures, a loss of fresh-water flow because of drought, and increased salinity as the sea level inches upward. Other species that spend most of their lives in the sea, such as flounder, sea trout, and red drum, come to the estuaries to spawn, where a warm greeting could translate into drastically reduced stocks. On the Pacific Coast, where there are far fewer estuaries, the loss of a San Francisco Bay to global warming would have profound consequences.

The implications of higher water temperatures are just as significant for inland fisheries. In great rivers like the Missis-sippi and the Missouri, fish can migrate northward into cooler habitats. But lesser rivers and most lakes do not cover as wide a temperature range, which would render them uninhabitable for a number of species if the waters continue to heat up. A 1995 study conducted by the Environmental Protection Agency suggests that the overall diversity of fishing in U.S. rivers and streams is likely to decline. Of fourteen cold- and cool-water species examined, all but one would have to retreat from at least one state if temperatures warm by two degrees Fahrenheit; and seven of ten warm-water species would also find at least one state too hot for survival. High water temper-atures also lead to lower levels of the dissolved oxygen that fish require. Another study estimates that if the waters in the rivers of the Southeast warm by seven degrees Fahrenheit, most of the native species will perish.

For several years fishermen living along the Atlantic Seaboard have watched in puzzlement as the populations of crab, oysters, and other coastal species decline, some because of overfishing or pollution, others owing to more vague and subtle forces that seem beyond individual control. On the scattered islands of the Chesapeake Bay, whose people have fished the great estuary for generations, sleep comes fitfully amid the sound of the lapping waters that daily erode their tiny ancestral domains. Little do they know that Teunaia Abeta slumbers just as fitfully an ocean away.

~

The most direct effect of climate change would be the impact of warmer temperatures themselves. Statistics on mortality and hospital admissions reveal that death rates rise during extremely hot days, particularly among the very old and the very young living in cities. People with heart problems are vulnerable because the cardiovascular system must work harder to keep the body cooler during hot weather.

The natural layer of ozone in the upper atmosphere blocks harmful ultraviolet radiation from reaching Earth's surface, but when concentrated at ground level ozone is a dangerous pollutant. Elevated temperatures increase the amount of the gas, damaging lung tissue and jeopardizing the welfare of those suffering from respiratory diseases such as asthma. Even modest exposure to ozone can cause an otherwise healthy individual to suffer chest pains, nausea, and lung congestion. In much of the United States, a warming of just two degrees Fahrenheit can increase ozone concentrations by about 5 percent, a figure that rises in the tropics.

In any given year, fires cover between 2 and 5 percent of Earth's land area, and humans ignite more than 95 percent of them. Tall-grass savannas, rain forests, and farms around the world are continuously ablaze. Not only does biomass burning contribute to the greenhouse effect by releasing billions of tons of carbon into the atmosphere, it has raised ozone levels in parts of Asia and Africa to values known to be toxic to both humans and plants. Satellite data indicate that ozone from fires in Africa travels clear across the South Atlantic and can be measured in easternmost Brazil.

This danger was recently brought home in a graphic manner to the citizens of Mexico, as well as Americans living along the Gulf coast. In the spring of 1998, some 13,000 blazes, many of them the size of a football field and spreading underground, began burning out of control during what should have been the rainy season in central and southern Mexico. Claiming many lives, the fires forced the closing of airports and left at least 50 million Mexicans choking in smoke. A cloud of haze and cinders covering thousands of square miles hung over the country for months, compounded by more uncontrollable burning in Guatemala, Honduras, and Nicaragua, another of El Niño's unwelcome gifts. While residents of Mexico City were placed under an environmental alert, the vast bank of smoke drifted northward across the Gulf of Mexico, blanketing states from Texas to Florida and creating a tickling in the throat, itchy eyes, and a wheezing in the lungs. With memories of Indonesia and Malaysia still fresh in their minds, Texas officials issued a health alert for the entire state, cautioning anyone with respiratory problems to stay indoors and everybody else, especially children, to avoid prolonged exercise.

South of the border, bewilderment gave way to despair. "The custom of our people is to go into the rain forest every year and bring down the trees," said Anselmo Entzin, a disconsolate resident of Chanal, a Tzeltal Indian town in Mexico's southern Chiapas state. "It hasn't rained since November, and when we went to put out the fires, we couldn't do it. This year our Lord in heaven put us in a bad problem."

These sentiments were shared by Kevin Trenberth, although he was reluctant to place the blame on divine providence. The head of the climate analysis section at NCAR believes that it is reasonable to look upon events like the haze in Texas as a "warning sign" of what the world at large might experience because of the greenhouse effect. El Niño can be viewed as a kind of template for global warming. "These effects are going to become more pervasive with climate change. Drying is likely to become more common, drought more severe and longer lasting, and this will make vulnerability to fire more pronounced."

Despite the statewide health alert, organizers of a twenty-six-mile relay race in Corpus Christi did not want to disappoint the 7,000 participants and so decided not to call it off. And like many others committed to regular exercise, a middle-age industrial engineer and his wife went for a jog along the trails of Houston's Memorial Park. Both wore blue masks to filter the polluted air, although health agencies made it known that this would do little good against minute particles that carry toxic and carcinogenic compounds deep into the lungs. When the engineer was approached by a reporter seeking his opinion on the possibility that a larger environmental issue was at stake, he paused just long enough to reply that the fears of global warming are part of a "Chicken Little Mentality."

~

The much-denigrated chicken was also on the mind of Rhys Jones, an archaeologist at the Australian National University in Canberra, as he was being interviewed by a reporter from the *New Scientist*. When queried about the ability of humans to adapt to the greenhouse world, Jones replied, "In the past, people experienced massive climatic changes on a scale that makes the greenhouse effect look like chicken feed." Yet the Aboriginal peoples survived and continued to live in much the same way as they had for tens of thousands of years.

Having said that, Jones hastened to qualify his remarks. While humans in the aggregate continued to live in much the same way for millennia, local populations were severely affected. Many groups abandoned their traditional living sites to ferocious cold or dry winds that blew across the land, carrying dust from the central deserts to the edges of the continent. A hundred centuries later, the more culturally advanced Anasazi and Greenland Vikings would do the same thing to avert annihilation, sacrificing all they possessed to the inexorable elements. Only the major civilizations, which had put down roots during the relatively benign postglacial age, were able to sustain themselves against floods, droughts, plagues, and insects, often at a terrible cost in human suffering.

Critics of the global warming theory often point to the historical record to bolster the argument that even if the planet is heating up, a concept that the most stubborn among them reject out of hand, a warmer Earth presents no significant threat to human survival. Yet no one on the other side of the debate is postulating the extinction of *Homo sapiens* as a species. Rather, it is a question of who will suffer the most and who will suffer the least if the predictions of the climate modelers come to pass.

Foremost among the skeptics is S. Fred Singer, professor emeritus of environmental science at the University of Virginia and a pioneer in the development of rocket and satellite technology. In *Hot Talk, Cold Science: Global Warming's Unfinished Debate*, the most recent of his fourteen books, Singer presents the dissenting view in a nutshell.

Singer is more impressed by what scientists do not know about global warming than what they have succeeded in proving to date, or what he refers to as science that is not "settled." To begin with, the resident time of carbon dioxide in the atmosphere is open to question. This is due in part to the fact that no one knows how much of it is being absorbed by the oceans, the elusive carbon sink. Some postulate an eightfold growth in the amount of atmospheric CO_2; others doubt whether the CO_2 level will so much as double in the twenty-first century.

The temperature record of the past hundred years is also dubious, largely because of the poor quality of the measurements. Ship observations are particularly questionable, while more recent readings from balloons and satellites disagree with those taken at Earth's surface. (This was before the publication of the Wentz-Schnabel finding.) The urban heat island, composed of cities and their sprawling suburbs, may also skew the record by artificially boosting readings.

Nor are Singer and his like-minded peers impressed by the results of the General Circulation Models currently in use. Some GCM temperature forecasts vary by as much as 300 percent and require arbitrary adjustments to bring them back into sync with reality. Their impossibly large grids allow major weather events to slip through undetected, and there is no satisfactory means of factoring in the microscale cloud processes that play a crucial role in determining temperature. Not only are the GCMs often out of tune with ongoing climate change, they cannot account for the unusual temperature rise from the early part of the twentieth century to the 1930s, or the cooling from around 1940 to 1976.

Who is to say that the climatic pattern will not shift just as abruptly in the future as it has in the past? Data from tree rings, ice cores, and sediments all tell of prehistoric climate fluctuations on time scales as brief as a decade, some of which have been greater than any changes forecast for the coming decades. The most recent El Niño may be just such an event, neither foretold nor accounted for by any GCM.

The limitations of the data notwithstanding, there is little doubt that global temperatures have increased in the past century. A number of scientists, and Singer is one of them, consider this warming a natural climate variation, most likely a recovery from the prolonged cooling of the Little Ice Age that ended during the reign of Queen Victoria. Why this "recovery" should continue at a record pace as yet another century draws to a close, Singer does not say.

Those who claim that a high CO_2 level is a recipe for global disaster, sincere though they may be, are badly misguided from Singer's perspective. In the long run, a higher CO_2 level "may turn out to be preferable to a lower one." This is the same argument advanced by Svante Arrhenius a century and more ago, when the Founding Father of global warming estimated that it would take another 3,000 years of burning fossil fuels to double the amount of CO_2 in the atmosphere. While Arrhenius was badly mistaken on that score, his concept of global warming as a benevolent phenomenon has not been completely dismissed by the IPCC or, for that matter, any other group of responsible scientists.

Experts who argue that warmer is better point to the fact that the number of people who die from the effects of cold weather will decrease. In the United States, freezing or stress due to low temperatures claims about 1,000 lives every year. However, the EPA reports that for every individual death attributable to cold, two others will die because of exposure to high temperatures. And deaths induced by the heat are more sensitive to temperature changes than deaths from the cold. For example, the difference between $-2°$ and $-15°F$ has a much smaller impact on mortality rates than an increase from $95°$ to $100°F$. In the heavily populated tropics and subtropics, where heat is the only temperature extreme that kills, the number of deaths from even more heat are certain to multiply.

With rising temperatures come undesirable insects, extending the range of disease vectors. While they acknowledge that this is a potential problem, the defenders of warming contend that a combination of better medical knowledge and more effective insect control technology will render the malaria-carrying mosquito and other harmful insect species innocuous. Whether public health officials would enjoy equal success in controlling rats, mice, cockroaches, and dust mites, the main suspects in the rapid increase in childhood asthma in many temperate climates, is problematic.

Projecting into the future, Arrhenius wrote that global agriculture would benefit from rising CO_2 levels, which may already be the case as the atmospheric fertilizer stimulates plant growth. More CO_2 also increases the efficiency with which plants use water, and may tend to offset some of the adverse effects of drier soils. As farmers acclimate to the new order of things, the mix of crops and livestock in a given region will be adjusted to take into account the altered growing season, temperature change, and the availability of water.

In the northern latitudes warmer nighttime temperatures, particularly in the spring and fall, will lead to longer growing seasons, which should contribute to agricultural productivity. It will also make it profitable to cultivate new lands in parts of Canada, Russia, Finland, Australia, and even Iceland. The cost will come when tractor and plow destroy ecosystems that have abided since the withdrawal of the glaciers. Meanwhile, the world's poorest nations are in the regions at greatest risk of projected crop losses—south Asia, tropical Latin America, and sub-Sahara Africa, where the expanding desert is claiming land rather than yielding it to the plow.

Warmer winters will reduce the demand for energy and with it the consumption of fossil fuels. A decline in the seasonal torments of snow and ice will make for safer driving. But as record temperatures arrive earlier and stay longer with the passing decades, such benefits will be offset as citizens of the industrial world become permanent hostages to air-conditioning, moving from home to car to office and back again with the thermostat set at a constant 72°F.

Fewer northerners will feel the need to escape to Florida or the Caribbean, not because of a rise in sea level but because of a more hospitable environment at home. Singer hypothesizes that the increased evaporation of the oceans will cause a more rapid accumulation of polar ice. If this scenario unfolds, the much-feared waters will actually recede, never more to haunt the dreams of Teunaia Abeta or those of his children's children.

More water vapor will also mean more clouds. Singer concurs with Richard Lindzen that new cloud formation, instead of trapping the infrared radiation coming from Earth's surface, will result in a negative feedback. The increased albedo will reflect more of the Sun's rays before they can reach the planet, canceling out much, if not most, of the anticipated warming.

As for life in the sea, warmer temperatures are likely to enhance fishing in many areas because overall biological activity is greater at higher temperatures. More food is available, fish grow larger faster, and they reproduce at an earlier age. Conditions will be particularly favorable for species that inhabit the high seas, while for species that live and breed near land it might be a very different story. Along the coast of Peru, perhaps the richest fishing ground on the planet, the anchovy and other species vanished in the wake of El Niño. Elsewhere, to the north, tuna, marlin, swordfish, and other species thrived as never before.

~

Thus does the debate continue, with no quarter given and none asked. Whether or not they are willing to admit it, participants on both sides of these frustratingly complex issues are filtering their arguments through their own ideologies, even as they claim science as their ally. The catch-22 of the greenhouse enigma is that there is still not enough scientific proof to fully predict and prepare for the even greater warming that may already be encoded on Earth's fragile shield. Yet to do nothing while awaiting further enlightenment may prove disastrous for the generations to come.

"It is simply hubris to believe that *Homo sapiens* can significantly affect temperatures, rainfall, and winds," writes Thomas Gale Moore of the Hoover Institute, a conservative think tank at Stanford University. "Global change is inevitable—warmer is better, richer is healthier."

Former secretary of the navy James Webb stands at the opposite end of the spectrum and favors a proactive course, believing that "we can (and do)" irreparably alter the environment by our excesses and lack of vision: "The wolf is in the door. Will we pretend not to see him . . . not to feel his breath . . . not to hear his footsteps? Listen to what our world is telling us."

This much we do know: No matter which way the dice fall, it is science and not personal conviction that will have the last word.

19

KYOTO

~

Nature never deceives us: it is always we
who deceive ourselves.

—Jean-Jacques Rousseau, *Emile*

I am not an Athenian nor a Greek,
but a citizen of the world.

—Socrates, quoted in Plutarch, *Of Banishment*

It is ironic that fifty-two years before hosting the 1997 United
Nations Conference on Climate Change, the city of Kyoto had
barely missed being destroyed. It was one of four cities being
considered as primary targets by President Harry Truman's
secretary of war, Henry L. Stimson, and General Leslie Groves,
head of the successful A-bomb project at Los Alamos, New
Mexico. The others were Kokura, Hiroshima, and Niigata.

Designated the site of a new capital by the emperor
Kammu in 794, Kyoto was laid out in the manner of Changan,
the capital of China's Tang dynasty. It served as the seat of the
emperors for more than 1,000 years until the Imperial Household moved to Tokyo in 1868, after the Meiji Restoration.
Nevertheless, Japan's rulers were still enthroned in the former
imperial palace, and Kyoto remained the center of Japanese
culture. Buddhist temples and Shinto shrines, which together
number more than 2,000, dominate the urban landscape. Inside the walls of these sacred structures are Japan's most important works of art: paintings, carvings, exquisite silks, fine
porcelain, cloisonné, and masterful examples of calligraphy.
Japanese theater was founded in Kyoto, and the city is surpassed only by Tokyo in the number of its institutions of

higher learning. All Japanese try to visit the city at least once in their lives, and nearly a third of the country's population make the pilgrimage each year. In the end, it was Stimson who recoiled at the thought of reducing all that beauty to atomic kindling, thus sparing Kyoto from the thunder roll of Armageddon.

There was a touch of the apocalyptic in the air as nearly 6,000 United Nation delegates from more than 160 countries began arriving in late November of 1997. The ten-day Conference on Climate Change was scheduled to open on December 1. In tacit recognition of the high stakes involved, most representatives from developing countries left their traditional dress at home, opting for the navy, black, and gray fabrics of diplomacy. Besides the delegates themselves, 3,600 representatives of environmental groups and industry were making the trek, together with nearly 3,500 reporters. Most were expected to crowd into Kyoto's International Conference Center, transforming its hallways into a modern Babel. And while the subject was climate, the forecast was for storms.

As the planet's largest consumer of fossil fuels, the United States, whose industry is responsible for some 35 percent of all the greenhouse gases ever created by human activity, was already cast as the villain. This role had recently taken on a darker dimension when the Clinton administration announced that it wanted emissions stabilized at 1990 levels. It further demanded that the world's industrial powers be given another ten years to achieve even that limited goal.

Hunkering down, Undersecretary of State Stuart Eizenstat, head of the sixty-member U.S. delegation, let it be known in advance that no amount of international pressure could force the administration to blink: "We want an agreement, but we are not going [to Kyoto] for an agreement at any cost." No sooner had the conference opened than Vice President Al Gore, an environmentalist of long standing, sounded an even more ominous note: "We are perfectly prepared to walk away from an agreement that we don't think will work." Anxious diplomats who had put the odds of drafting a treaty at fifty-fifty were suddenly hedging their bets.

Unruffled by this tough talk, German foreign minister Klaus Kinkel took the United States to task. With a poker-faced Eizenstat and the rest of the U.S. delegation looking on in silence, Kinkel reminded the Americans of their historic tradition as innovators: "Pioneers would be expected to set high standards. Future generations must not be burdened with the costs of our carelessness." As the spokesperson for the European Union, the foreign minister proposed a 15 percent reduction in greenhouse gases from their 1990 levels by 2010.

When Britain's turn came, Michael Meacher, the environment minister, upped the ante. While the British government appreciated the difficulties U.S. officials faced in battling their aggressive industrial lobby, "they certainly ought to be making, in our view, a bigger effort." Meacher proposed that carbon dioxide emissions from transport and industry be cut by 20 percent of their 1990 levels by 2010. Even so, the minister was fully aware that such a reduction would not be nearly enough in the long run. The most advanced computer models were projecting that greenhouse emissions would have to be lowered by about 70 percent in order to negate global warming.

Tuiloma Neroni Slade, chairman of the Alliance of Small Island States, welcomed both Kinkel's and Meacher's remarks. His group of forty-two countries could one day disappear under rising seas if temperatures continued their upward spiral. Working feverishly behind the scenes, Slade was lobbying for emission cuts even tougher than those proposed by some environmental groups, well knowing that until there is a political transformation in rich and poor nations alike, the prospects for substantive change are remote.

In the wake of this pummeling, the remarks of Hiroshi Oki, Japan's environment minister and president of the Kyoto summit, seemed mild by comparison. After urging flexibility on the part of the United States, Oki rejected the substantial cuts in carbon dioxide emissions recommended by Britain, the European Union, and others. Instead, the Japanese government, itself under intensive pressure from industry, would accept a 5 percent reduction in 1990 emission levels, provided it was phased in over the next twelve years.

In contrast to the diplomatic finger-pointing and grueling behind-the-scenes negotiations, the atmosphere outside the conference center bordered on the playful. In the parking lot,

Greenpeace assembled a giant solar array, using it to power an environmentally correct kitchen, with an electric stove able to fry an egg. Visitors to the exhibit were offered tea and mocha brewed with solar energy and departed with handfuls of pamphlets printed on recycled paper.

Environmentalists who normally couldn't get their phone calls returned were suddenly the focus of the world's media, and they reacted like newly discovered rock stars. Malini Mehra, a member of Friends of the Earth, compared the atmosphere to "a circus." "It's just a jamboree. Adrenaline is just racing." Greenpeace member Kalee Krieder gave interviews back-to-back all morning on the opening day of the conference. "There's ten minutes of fame in your life," she gushed, doing Andy Warhol one better. "Here it is. I was born in Zanesville, Ohio, and I never thought my day would come in Kyoto, Japan."

Australia, one of several countries that sought and ultimately won special concessions, was lambasted at a well-attended news conference by an activist wearing a diplomat's gray suit, a blue tie, and brown paper sack over his head. "It's very embarrassing that Australia has dealt itself out of these negotiations by taking such a maverick, recalcitrant position," said Michael Rae of the World Wildlife Fund after snatching off the bag with a dramatic flourish.

As silly as much of this may have seemed, these were the direct heirs of the student activists of the 1960s and 1970s, who were largely responsible for sparking the environmental movement even as they marched against the Vietnam War and demonstrated for civil rights. Although they were now middle-aged and scattered to the winds, Kyoto belonged to them as much as to anybody. Following these developments from afar, they were finally experiencing what it was like to be fenced in rather than fenced out.

With about forty of its members in attendance, the largest international environmental delegation belonged to Greenpeace. Still, it was no match for the opposition when it came to numbers and economic resources. Hundreds of industrial lobbyists had poured into Kyoto and were doing all they could to derail the kind of comprehensive treaty that environmentalists had been touting for years. Among the most powerful of the industrial associations was the Global Climate Coalition, whose name had a deceptively environmentalist ring. Representing many of the largest U.S. corporations, including Exxon, Mobil, and Shell Oil, the Big Three automakers, mining and transport companies, steelmakers, and chemical producers, the coalition had already launched an aggressive $13 million advertising campaign. Television viewers were warned that strict reductions in greenhouse gases would have catastrophic economic consequences, endangering the lifestyle of every American. Gasoline would shoot up by fifty cents or more a gallon; heating and electricity bills would soar, while higher energy costs would raise the price of almost everything Americans buy. The livelihood of thousands of coal miners, autoworkers, and others employed in energy-related fields was on the line.

The coalition also had a ready answer for those concerned about the price to be paid for doing nothing. In another, more benign, series of ads, the narrator assures the viewer that in a warmer world crops will continue to grow more productively. Forests will expand their ranges. Grasses will flourish where none took root before, and great tracts of barren land will be reclaimed.

Yet according to William F. O'Keefe, the Global Climate Coalition's chairman, the debate had gotten ahead of itself. In an interview with Peter Jennings of ABC News, O'Keefe seized on the issue of scientific uncertainty, taking a position that was in direct conflict with the coalition's self-assured media campaign: "Scientists have said that we do not understand the climate system that well. We don't understand the role of oceans, water vapor and clouds, and that to really make these long-term projections, we need to have a better understanding of the system." When pressed by Jennings to explain why it is that the industries considered most responsible for global warming are the most critical of its scientific validity, O'Keefe replied, "First of all, that presumes that industries are causing global warming. Carbon dioxide comes from people exhaling, from decaying plants. It comes from a number of natural sources. . . . Human activities, which include burning fossil fuels, account for a very small percentage."

Hoping to boost their credibility, coalition members had

been recruiting scientists of their own to study the issue. Among them is Robert Balling, a climatologist at the University of Arizona. "I'll get to my grave," Balling quipped to Jennings, "and people, I'm convinced, will still be debating whether or not the buildup of the greenhouse gases has had much of an impact on global climate." As for accepting funding from a company seeking to protect its vested interests: "The coal company may pick a project because they think that project may provide a set of answers that the coal company would be happy with. But the coal company can't come to my laboratory and sit there and take the printouts and say, 'OK, take this one, not this one.'"

When given the chance to respond, Vice President Gore cited the government's long struggle with the tobacco industry, whose hired guns had procured a large amount of "junk science" with which to refute the charge that cigarette smoking causes lung cancer. "I think that it is totally fair and totally legitimate to say that we're going through the same kind of experience again. Remember how wrong they were? Remember what their arguments sounded like? Now listen again to what these people are saying."

Gore was just as adamant on the issue of economics and global warming. "Every time in our history we've tried to clean up pollution, the doomsayers have said this is going to cause an economic catastrophe, and every time they've been wrong." But what about the price, Jennings wanted to know, of gasoline and electricity, 50 percent of which comes from coal? "Right now," Gore countered, "two-thirds of all the energy that we burn to produce electricity is completely wasted. It's turned into pollution, not useful energy. With competition and new technologies, that's going to change." Not only will energy prices remain low, according to Gore, but "utility bills can come down, and will come down."

Their superior numbers and large war chest aside, the lobbyists proved no match for the environmentalists on the propaganda front. Friends of the Earth set up a ballot box to solicit votes for the "top Dirty Dozen" companies lobbying the conference. Dressed like insurgent commandos in khakis, boots, and dark sweaters, they held forth with stacks of statistics and research papers, buoyed by the knowledge that local hotels were asking guests to reuse their towels to cut down on laundry waste while hundreds of young Japanese guides greeted delegates wearing bright green jackets. In the background, video screens continuously warned about the dangers of climate change. At one news conference after another, industrial lobbyists were blasted for their "obstructive role" and for "unabashedly playing games with science and statistics."

So intense was the onslaught that these staid professionals, who had honed their negotiating skills working quietly behind the scenes in the corridors of power, were feeling like outcasts. J. R. Spradley, a Washington lawyer and representative of the capital's Edison Electric Institute, had come to deliver the message that Kyoto threatens to "wreck the entire world economy." Instead Spradley wound up taking solace from the Old Testament and a fractured view of history: "How many people were following Moses when he started? And there was only one guy saying the Earth was round in the beginning; it's nothing to be ashamed of." Gail McDonald, former chairman of the Interstate Commerce Commission in the Clinton administration, was one of those singled out by Friends of the Earth for her lobbying on behalf of the Global Climate Coalition. "For a person who used to get a certain degree of deference as a public official, it's hard," McDonald confided to a reporter from the *Washington Post*. "I remind myself each day that I need a thicker skin."

~

Of all the political leaders to attend the Kyoto Summit, few, if any, possessed more solid credentials as an environmentalist than Albert Gore Jr. In 1992, while he was on the threshold of becoming vice president of the United States, then-Senator Gore published his paean to global ecology, *Earth in the Balance*. His consciousness about global warming had been awakened when he signed up for an undergraduate course on population in the mid-1960s. The professor was Roger Revelle of Scripps, who shared with his students the dramatic results of Keeling's measurements of rising carbon dioxide concentrations in the atmosphere. "The implications of [Revelle's] words were startling," Gore wrote. "We were looking at only eight years of information, but if this trend continued, human civilization would be forcing a profound and disruptive change in the entire global climate."

Gore later went to Vietnam as a military reporter with the army and wrote of jungle transformed into a stark imitation of the lunar landscape by a herbicide called Agent Orange. After his election to Congress, he became expert on a recently discovered waste dump at an obscure place in upstate New York called Love Canal. When the ozone hole opened in the skies above Antarctica, Gore put his political campaign on hold and drafted a major speech outlining a strategy for eliminating chlorofluorocarbons. Expecting front-page coverage, the politician said he was crushed when "not a single word" was written in any newspaper in the United States about the speech.

Gore persevered, and by the time George Bush became president the fossil fuel industry had organized to fight the senator's ecological crusade. William Reilly, former administrator of the Environmental Protection Agency, credited the efforts of the National Association of Manufacturers, the U.S. Chamber of Commerce, certain unions, and the chemical, auto, aluminum, coal, and steel industries with forcing the president to knuckle under, giving "politics and economics primacy over the issue of science at the time." All of a sudden, global warming wasn't science, it was science fiction.

Gore retaliated by inviting James Hansen to testify a second time before the Senate Committee on Energy and Natural Resources, in 1989, a year after the government scientist's bombshell testimony had alerted many Americans to the dangers of global warming, prompting calls for action. Put on notice, the White House, operating through the Office of Management and Budget, forced Hansen to alter his prepared statement. Having described the connection between warmer temperatures and more frequent droughts as a "probable" consequence of global warming, a conclusion based on his scientific research, Hansen was forced to change the language to "highly speculative."

A livid Gore charged the administration with "science by fraud" and compared the actions of the White House to those of the Soviet Union, where scientists were coerced into shaping their studies to fit prevailing ideology.

Three years later, in June of 1992, the countries of the world gathered in Rio de Janeiro, Brazil, for the United Nations Conference on Environment and Development, otherwise known as the first Earth Summit. The overwhelming majority, led by the Europeans, favored some kind of mandatory limits on greenhouse gas emissions, a step opposed by President Bush, who succeeded in holding out until the limits were made voluntary in a treaty written in ambiguous and convoluted language. Tensions were further heightened when the United States refused to join the 153 other nations in endorsing a second treaty on biodiversity, committing the signatories to the protection of the world's plant and animal species. Before Bush left Rio, he told a reporter from a local daily, "I'm President of the U.S., not President of the World. It is leadership to stand up against the tide. I can't do what everyone else does." Air Force One was just climbing into the afternoon sky on the long journey home when members of Greenpeace scaled Sugarloaf Mountain and draped Rio's landmark with a large banner denouncing the Earth Summit for having "sold out" the planet.

Two months after Rio, Arkansas governor and presidential candidate Bill Clinton chose Gore to be his running mate. Their victory in November gave the new vice president a powerful platform from which to resume his crusade against human-induced climate change. He soon won Clinton over, and in October of 1993, less than a year after their election, the two issued "The Climate Change Action Plan." At Rio, the United States had joined the other participants in signing the Framework Convention on Climate Change, whose main objective was the stabilization of greenhouse gas concentrations in the atmosphere. The new administration committed itself to returning U.S. emissions to 1990 levels by the year 2000, at the same time creating jobs and stimulating investment. Gore's signature could be read in the action plan's every line; his star was riding high.

Still, writing a book and drafting position papers was one thing; retaining the support of environmentalists while not antagonizing business leaders and union membership was quite another for a man harboring presidential ambitions. The trouble began for Gore when the time came for the Clinton administration to lay out its position on the upcoming negotiations in Koyto.

In early October of 1997, the CEOs of the Big Three U.S. automakers met with President Clinton to voice their opposition to a global warming treaty. They had already begun running TV ads opposing any such document, one of which showed a map of the world with scissors cutting out those countries that would be exempt. What was more, United Auto Workers president Stephen Yokich supported the ad campaign, an ominous sign for any Democrat.

As industry and labor leaders well knew, the issue of exemptions was the administration's Achilles' heel. At Rio, the industrial countries had pledged to voluntarily reduce greenhouse gas emissions to their 1990 levels by 2000, and it had not worked. Instead, even more carbon dioxide, methane, and other trace gases were being spewed into the atmosphere. As a rueful President Clinton told a reporter, "Regrettably, most of us, including especially the United States, fell short." To reach that failed goal, it would now take a 20 percent reduction in emissions by 2010.

At the same time, Clinton and Gore had made it clear that they were sympathetic to the plight of the world's poorer nations. They backed the stance taken by Bert Bolin, the Swedish climate scientist who had recently stepped down after nine years as chairman of the IPCC. While calling on the industrialized nations to establish mandatory limits on greenhouse gas emissions at Kyoto, Bolin was sharply critical of any effort to make the underdeveloped nations do the same, an exemption that had been granted them at Rio.

The White House's position on this issue had already been dealt a body blow. In July, almost five months before Kyoto, Senator Chuck Hagel, a Nebraska Republican, introduced a nonbinding resolution urging the administration not to pursue a treaty unless the agreement required developing nations to control their emissions as well. The resulting vote was 95 to 0, putting the president and vice president on notice that the likelihood of their ever getting the Senate to ratify a treaty without such a clause was virtually nil. While senators fretted over the serious harm a "one-sided" agreement would do to the U.S. economy and its competitive position globally, many environmentalists expressed support for the resolution as well. Sympathetic as they were to the plight of the underdeveloped nations, they were deeply concerned about recent scientific projections. By 2015 or so, the nations of Asia, led by China and India, would become the world leaders in the release of carbon dioxide. The bitter legacy of colonialism notwithstanding, to let them off the hook could prove disastrous for the world.

Clinton, and more particularly Gore, were now forced to backtrack. In 1995 they had supported what United Nations officials called the "Berlin Mandate," which would exempt 134 developing nations, including China, India, and Mexico, from whatever limits on greenhouse gases were hammered out in Kyoto. This miscalculation handed industry and the political opposition a club with which to pummel the administration. Clinton and Gore finally announced that the United States would scuttle its prior commitment and insist that the developing world participate, prompting charges of betrayal on the eve of Kyoto.

Thus the scene was set whereby the politically powerless, weary of their hardscrabble existence, would take turns lecturing the rich and the strong, with the Americans bearing the brunt of their ire. And just to be sure that negotiators did not stray from the straight and narrow, Senator Hagel was heading a congressional delegation at Kyoto, with whom U.S. diplomats were pledged to keep in close contact.

As for Gore, his star had seemingly become a meteorite. A day into the Kyoto Summit, Friends of the Earth began distributing a leaflet portraying a jowly man in a cowboy hat and bolo string tie, resembling the television character J. R. Ewing of *Dallas.* The bloated figure is the caricature of big oil and big automotive companies that "are making billions of dollars from fossil fuels whilst helping to wreck the Earth's climate," as the leaflet put it. Seated on the Texan's considerable lap are two diminutive puppets, one representing President Clinton, the other Vice President Gore.

~

By day four of Kyoto an editorial in the *New York Times* declared that a "near-miracle" was required for the summit to be salvaged. Lobbyists on both sides had left U.S. negotiators with little room to maneuver, as evidenced by the unfolding of a bizarre nightly ritual. First, one hundred or so lobbyists representing the business community would gather in a marble-lined room to be briefed by members of the negotiating team. When their half hour was up, the lobbyists were ushered out, and representatives of the environmental community would take their place and receive the same briefing. As the talks ground on, everyone agreed that the issue most likely to block a compromise was the degree to which developing countries should participate in curbing global warming. Richard L. Trumka, former president of the United Mine Workers, stated that if any countries are exempted from emission-control targets, "we think our Government negotiators need to be prepared to walk away from Kyoto without a treaty in hand." Speaking for many of the environmentalists, the Natural Resources Defense Council took a less strident position: "The U.S. and other industrialized countries must participate in efforts to limit emissions. But the U.S. must not evade its own responsibilities by making unrealistic demands of poor countries."

Fearing that Kyoto would collapse and that the United States would be blamed, President Clinton dispatched Gore to Japan in hopes of luring delegates from all sides to the middle ground. Not wanting to become an issue himself, Gore stayed in the ancient capital a brief eight hours, during which he addressed the conference, met with environmental groups, and spoke with members of Congress.

The whirlwind visit paid off. Following Gore's speech, in which he promised greater flexibility on the part of the U.S. delegation, diplomats from Europe, Japan, and the United States were said to be meeting quietly, perhaps to work out an agreement that would be announced soon. Still, the squabbling continued. Shortly after the vice president departed the conference hall, diplomats entered into a heated exchange on a subject they had already discussed ad nauseam: how many greenhouse gases the treaty should limit.

Meanwhile, world leaders were on the phone to one another in an effort to resolve this conflict and others. In the last few days of the summit, President Clinton spoke to Russian president Boris Yeltsin, British prime minister Tony Blair,

German prime minister Helmut Kohl, and Japanese prime minster Ryutaro Hashimoto.

On Wednesday, December 10, weary but still hopeful delegates awoke to the devastating news that no agreement had been struck. By midnight, Kyoto would have become part of history. Then came word from conference chairman Raul Estrada of Argentina that he and others had worked through the night, and that there had been mcvement toward making the draft treaty more strict. "The trend is a very positive one," he said.

Choosing his words with great care so as not to kindle false expectations, chief U.S. negotiator Stuart Eizenstat told the press: "We still have far to go. Nevertheless, we are hopeful that . . . we will be able to bridge the gaps."

The final day stretched beyond the midnight hour and into the next morning. The future of the carbon-charged atmosphere now hung on the benign click of calculators as negotiators in back rooms set budget periods, apportioned quotas, and added up projected emissions of gases linked to global warming.

The rumor was true after all: The Europeans, the Japanese, and the Americans were negotiating in secret. In the small hours of December 11, a visibly exhausted Raul Estrada alerted the press that an agreement had been reached. Minutes later came the announcement. The United Nations Conference on Climate Change had produced the first treaty containing mandatory reductions in greenhouse gas emissions. "If approved, the Kyoto Protocol will set an energy course for much of the world for decades to come."

Shaped by what the press was calling the "grand European-U.S.-Japanese compromise," the document calls for the European Union to reduce its greenhouse emissions by 8 percent below 1990 levels, the United States by 7 percent, and Japan by 6 percent between 2008 and 2012. Twenty-one other industrialized nations would have to meet a similarly binding target of 5.2 percent. Six gases would be affected: carbon dioxide, methane, nitrous oxide, and three halocarbons used as substitutes for ozone-destroying chlorofluorocarbons.

Precisely how these cutbacks are to be achieved the protocol does not say, but governments are expected to take steps such as converting coal-fired power plants to natural gas, encouraging the production and development of more fuel-efficient automobiles, and ending subsidies to keep fossil fuel prices artificially low.

Led by China, it was the developing countries that got what they most wanted. Facing the prospect of international condemnation and a backlash on the part of environmental groups, the United States agreed to exempt the developing countries from mandatory emission controls. In the language of the protocol, these nations can "opt in" under the binding agreements, though few were naive enough to believe that any of the significant polluters would do so. Also blocked was a U.S.-backed proposal that the industrial nations be permitted to trade "emissions quotas" among themselves. This would have allowed a country to fall short of meeting its cutback target by purchasing quotas from another that more than meets its target. In return for dropping this demand, the language of the protocol called for it and many other details to be worked out at two international conferences scheduled to take place in 1998.

Finally, the draft language calls for the protocol to take effect once it is ratified by at least fifty-five nations that collectively account for at least 55 percent of 1990 carbon dioxide emissions. The terms become binding on an individual country only after its government ratifies the treaty.

Although the green movement generally wished the emissions targets had been more ambitious and inclusive, most agreed with John Adams, director of the Natural Resources Defense Council, who characterized the steps taken at Kyoto as "an immensely important turning point." Philip E. Clapp, president of the National Environmental Trust in Washington, called the agreement "a historic landmark in environmental protection," one that will be remembered "as a central achievement of the Clinton-Gore Administration." Only the World Wildlife Fund took major exception, calling the outcome "a flawed agreement that will allow major polluters to continue emitting greenhouse gases through loopholes."

Feeling betrayed, industrial lobbyists and congressional opponents were much less generous. "It is a terrible deal and the President should not sign it," lamented William O'Keefe of the Global Climate Coalition. Neither was Senator Hagel about to mince words: "There is no way," he told the press before departing Kyoto, "if the President signs this, that the vote in the United States Senate will even be close. We will kill this bill!" Senate majority leader Trent Lott was only slightly more charitable, calling the treaty's prospects "bleak" before the ink was dry.

For his part, Al Gore seemed to relish the thought of doing battle with the Senate. "It will be a real knock-down, drag-out debate that would be great for the country."

CODA

~

This has been the biography of an idea, or rather, I should say, the partial biography, for its subject is very much alive and gives every indication of remaining so for a long time to come. How long, one wonders, will it take before the riddle of global warming is resolved beyond any reasonable doubt? Certainly not the 15 billion years that passed before the universe whispered to Copernicus that Earth is a planet circling a star, or the 4 billion years Earth took to reveal its ancient origins to Hutton and Lyell. As in times past, scientists find themselves in much the same position as historians who are asked to render judgment on the period in which they live. As Sir Francis Bacon observed long ago, we are dancers in the ring, unable to see the beginning, the middle, or the end. There are clues, connections to be made, but when will the moment of certitude come—tomorrow, next year, a decade hence? When?

As I write this page, it is six months since the Kyoto Summit. After an unusually mild and almost snowless winter, courtesy of El Niño, spring has come to Indiana. I feel the urge to plant something, perhaps a tree, and drive with my wife to the local nursery to take stock of the saplings. While she busies herself with the bedding plants, I stroll alone among the red maples, poplar, flowering crab, and pin oak before pausing beside a Canadian hemlock. As I stroke its cascading green branches, it occurs to me that something is not quite right. The sky, which only this morning was a vibrant blue, is now tinged a strange yellowish gray. Then I remember the weather forecast of the previous evening. The smoke from the thousands of uncontrolled brush fires in Mexico has migrated northward beyond the Gulf states and will reach the Midwest in a day or two—indeed has done so already. I am suddenly aware of the oppressive heat. Though it is only mid-May, the temperature is 96°F, a record for this date. The air is saturated by humidity from a couple of recent gully washers measuring two-plus inches, the kind scientists say will become more commonplace in a greenhouse world.

"Need any help?" I startle and turn around. The voice is that of a young man dressed in chinos and a T-shirt, who looks to be a high school student working on the weekends.

"No, thank you, I'm just looking. It's a bit on the warm side to be planting trees. Must be global warming; what do you think?"

He gives me a quizzical look, as if I am his science teacher and have asked him a trick question. He thrusts his hands deep into his pants pockets, and with a barely perceptible shrug turns and walks away, saying nothing.

Driving home with the air-conditioning on high and a trunk filled with flowers but no tree, I suddenly remember the large file folder on my study desk. It is bulging with magazine and newspaper clippings, all collected in the months following Kyoto. I begin thumbing through them in my mind.

For several months a young woman named Julia Hall—or "Butterfly" to her fellow members of Earth First!—has endured the wet and cold of an El Niño winter perched atop a wooden platform 180 feet above the ground. Her home is a California redwood she calls Luna, whose trunk has been marked with a paint slash by loggers who are waiting impatiently for her removal so they can fell the ancient giant. With electric socks to ward off the chill, a cell phone to carry on live radio interviews, and a propane stove to cook her Mexican and Chinese dinners, Butterfly is North America's new champion "tree sitter." Pointing to a large banner unfurled above her aerie that reads Respect Your Elders, she declares, "We have to stop the rape of the forest." Only five miles away, at the world's largest redwood mill, sixty-foot band saws are turning trees like Luna into $100,000 worth of lumber in two hours.

Butterfly is not as alone as some might think. According to another article, executives at American Electric Power, a large utility company, have purchased a Bolivian jungle, complete with howler monkeys, jaguars, 600 species of birds, and insects by the billions. The tract will soon become part of a national park and is expected to remove 58 million metric tons of carbon dioxide from the atmosphere over the next thirty years.

Together with American Electric, eleven major corporations have broken ranks by entering into an initiative against global warming. Whirlpool, 3M, Toyota, Sunoco, United Technologies, and Lockheed Martin are among the industrial giants joining the Pew Center for Global Climate Change, an arm of the Pew Charitable Trusts, the largest of the nation's pro-environment grant makers. A half-page ad in the *New York Times* announcing the agreement reads: "We accept the views of most scientists that enough is known about the science and environmental impacts of climate change for us to take actions to address the consequences. Kyoto ... represents a first step in the international process, but more must be done both to implement the market-based mechanisms that were adopted in principle ... and to more fully involve the rest of the world in the solution."

The saving and replanting of forests, or what scientists call the sequestration of carbon, is only one part of the equation, according to Dan Becker, the head of climate programs at the Sierra Club. Becker estimates that in order to offset annual industrial emissions from the United States alone, "you would need to plant a new forest on an area the size of Australia. You are taking carbon that is safely sequestered underground, as coal or oil, and bringing it up, and adding it to the atmosphere. And then you are temporarily storing it in a closet made of trees. I am all for preserving forests, and I am against cutting down forests. But is it a good thing to pollute more because you have done that? No." Michael Oppenheimer of the Environmental Defense Fund is also dubious: "If sequestration is done correctly, it can produce many benefits. It can enhance ecosystems and remove carbon dioxide from the atmosphere. But done poorly, it can make the greenhouse problem worse and do a lot of damage to the ecosystem." Brent Blackwelder, president of Friends of the Earth, believes the corporate emphasis on forestry is a dangerous distraction and wants companies to use their resources to develop cleaner technology and reduce their dependence on fossil fuels. "We've got to do a whole lot of rethinking about how we fuel the global economy," he says. "Planting trees is not sufficient."

Another ad in the *New York Times*, this one placed by the World Wildlife Fund, trumpets "A Victory for the Amazon!" Brazil's president, Fernando Henrique Cardoso, has just announced that he has committed to triple the area of the rain forest under government protection, an additional 62 million acres of new parks and preserves, greater in size than the entire U.S. National Park system outside Alaska. The pledge is to be carried out with financial and technical assistance from the World Bank and the World Wildlife Fund International.

If only it were so simple. Nigel Sizer, an expert on Amazonian forests and a senior associate at the World Resources Institute, considers the pledge important, but no panacea. "It will not by itself significantly reduce the rate of deforestation going on in the region, especially on the frontiers," he notes. "The root causes that continue to cause deforestation to accelerate are not addressed." It will take years, perhaps decades, to implement such a plan, and Brazil has never been strong when it comes to the enforcement of environmental law. If violators are punished at all, the most they pay is a modest fine.

In an article that appeared only three days before President Cardoso's announcement, the Brazilian government reported that the destruction of the rain forest had tripled between the 1990–91 and the 1994–95 burning seasons. Seven million acres, covering many thousands of square miles, were destroyed in the latter period alone, and this number does not take into account much of the acreage consumed in the intense and widespread fires triggered by El Niño. Already, 130 million acres of the Amazon's nearly 1 billion acres are deforested. Fearing an international outcry, Brazil postponed the release of these figures while attending the Kyoto Summit.

A few days before this announcement, *New York Times* reporter John H. Cushman Jr. broke a major story. Still licking the wounds they had sustained at Kyoto, the industrial opponents of the treaty had already drafted an ambitious proposal designed to convince the public that the environmental accord is based on shaky science. Meeting secretly in the Washington office of the American Petroleum Institute, employees of Chevron, Exxon, and the Southern Company, the trade associations, and conservative policy research organizations are prepared to spend millions more in support of their cause. Much of the budget is to be directed at lobbying science writers, editors, columnists, and newspaper correspondents in an effort to convince them that the risk of global warming is too uncertain to justify controls on greenhouse gases. Another part of the plan calls for the recruitment of a cadre of scientists who share industry's views on climate change and training them in public relations.

Among the participants is the physicist S. Fred Singer, author of *Hot Talk, Cold Science*, and the founder of the conservative Science and Environment Policy Project. Singer has been joined by his friend and colleague Frederick Seitz, who wrote the foreword to Singer's book.

One week before the Cushman article appeared, the National Academy of Sciences, of which Seitz was president in the 1960s, took the extraordinary step of disassociating itself from a statement and petition circulated by the retired physicist. The petition, which was sent to thousands of scientists, called for the government to reject the Kyoto Protocol. It was accompanied by what appeared to be the report of a scientific study concluding that carbon dioxide emissions pose no climatic threat but instead amount to "a wonderful and unexpected gift from the Industrial Revolution." The article was printed in a format and typeface similar to that of the academy's prestigious peer-review journal.

Few scientists who received the petition were climate experts, and many of the rest became suspicious when the document was traced to the obscure Oregon Institute of Science and Medicine, based in Cave Junction, population 1,125. Robert L. Park, a physics professor at the University of Maryland and the author of *Voodoo Science*, stated: "I don't know how many petition cards were sent out, but I can guess who paid for the mailing. There is a well-financed campaign by the petroleum industry to recruit scientists who are skeptical about global warming to help convert journalists, politicians, and the public to their views."

While the corporate heirs of Carnegie and Rockefeller battle on, so too do those of Henry Ford. Detroit is doing record business, spurred by the baby boomers' demand for so-called light trucks. Americans are now purchasing more sport utility vehicles, minivans, and pickups than passenger cars, and manufacturers predict that the introduction of nearly two dozen more new sport utility vehicle models over the next several years will lure ever more buyers. With women fast joining the rising tide, the race is on to build more factories and convert existing plants to the production of light trucks, which reap much higher profits than cars. It is estimated that these vehicles will account for 55 percent of family sales by 2003.

One week after the Kyoto Protocol was adopted at the midnight hour, the Environmental Protection Agency proposed new air-quality regulations that would not affect the already less stringent emissions rules for the big sport utility vehicles and pickups. The largest of the light trucks would be allowed to emit five and a half times more nitrogen oxides, the main ingredient of smog, than cars, while slightly smaller light trucks would be allowed three and a half times the emissions of automobiles. The ghost of Charles Erin Wilson, former president of GM and President Eisenhower's secretary of defense, still wanders the corridors of bureaucracy: "What is good for the country is good for General Motors, and what's good for General Motors is good for the country."

~

El Niño's most recent irruption has come to an end. Unlike a number of previous episodes that lingered for years, the Pacific warming that scrambled atmospheric patterns and caused billions of dollars in damage worldwide made a rapid exit. Meanwhile, the Child's coldhearted little sister, La Niña, is being reborn, and climatologists predict that she's likely to throw her own weather tantrums in the fall and winter of 1998, as well as the spring and summer of 1999. Early warnings are being posted for wind-driven brush fires in southern California, flooding in the Pacific Northwest, drought in the

Southwest, and more-powerful-than-average hurricanes in the Gulf of Mexico and the Atlantic. In China, where record flooding along the Yangtze has displaced millions and exhausted laborers are preparing for the river to crest for the fifth time in weeks, government officials are already pointing an accusatory finger at La Niña.

With climatic transition as a backdrop, the White House recently announced that government scientists have determined that temperatures in each of the past fifteen months have broken global highs for that month. July of 1998 averaged 61.7°F, 0.6 degree higher than July of 1997. Previously, such year-to-year changes were measured in hundredths of a degree and took a century to add up to a single degree Fahrenheit. As if to emphasize this point, in 1998 a heat wave slammed Texas with twenty-nine consecutive days of triple-digit temperatures and claimed 126 lives, not counting those who perished trying to slip undetected across the border from Mexico. Temperatures in Kuwait hit 122°, and Egyptians, long used to toiling in the desert, took to working at night after many heat-related deaths. The thermometer soared to the century mark in Paris, prompting citizens and tourists to seek relief in the fountains near the Eiffel Tower and the Louvre. Throughout Europe homes and businesses are being rewired to accommodate the previously unthinkable—air-conditioning. It's as if a specter from the future has suddenly materialized a century before its time, giving us a taste of what awaits us on the other side.

With the demise of El Niño, scientists thought it unlikely that the waning months of 1998 would be as warm as the period from January through September, though the year would easily set another temperature record. (Months later they were able to confirm that it had done so in what the British climatologist Philip D. Jones termed "amazing" fashion. The year was not only the warmest in the thermometer record, but was also the warmest year of the millenium. Moreover, it had topped 1997 by an unprecedented quarter of a degree, meeting yet another of James Hansen's litmus tests.)

The announcement of the climatologists' findings was left to Al Gore, who was hoping to send Congress a message that it is urgent to enact a major program of financial incentives and technological research aimed at cutting the emissions of greenhouse gases, a measure that would put the United States on track to live up to the Kyoto Protocol. But during the summer of 1998, Republicans on the House Energy and Water Development Appropriations Subcommittee voted to prohibit funding for programs to implement parts of the Kyoto agreement through presidential executive orders and federal departmental directives. House members were following the lead of their colleagues on the Senate Appropriations Committee, who had already approved language critical of administration plans to implement the Kyoto treaty before it is ratified. Thus do matters stand, although some Republicans seem to be having second thoughts. By rejecting the Kyoto treaty and the mounting scientific evidence that global warming is indeed at hand, they might be providing Vice President Gore or some other Democrat a free ticket to the White House, since polls show that up to two-thirds of the voting public are deeply concerned about the future of the environment.

~

Thanks to the improved quality of the air, a resident of Manchester, England, is now able to gaze on the distant Pennines for the first time in more than a century. In and around the city, *Biston betularia*, the peppered moth, thrives, but an interesting thing has happened. *Carbonaria*, the dark melanic type, is fast disappearing throughout its range, while its mottled cousin is mounting a dramatic return—so dramatic, in fact, that entomologists believe *carbonaria* could soon become as rare as it was during the early days of the industrial revolution, when the mystery of its origins haunted the bespectacled R. S. Edleston.

Now comes another, even deeper, mystery involving creatures so durable that they once sang to the dinosaurs of the Jurassic 150 million years ago. The frogs and toads of the world are dying, and experts are dumbfounded as to why. From Canada to Costa Rica, Yosemite to Yellowstone, Brazil to Indonesia there is silence in the forests of the night. Is it something humans are doing, and if so, is it something that will eventually threaten our species? Hypersensitive to changes in the environment, whether water or the air, these amphibians are the first to vanish when their habitat is threatened. One theory holds that it is the increase in the ultraviolet light from holes in the ozone that reduces the creatures' immunity to parasites, viruses, and fungi. Another maintains that it is the rising temperatures of the planet and pollution borne on water and wind. If extinction comes, this much we will know for certain: It is the end of fairy tales. For who could ever again tell a child that the frog who was changed into a prince and the beauty who bestowed that magic kiss lived happily ever after?

~

On November 12, 1998, the United States affixed its signature to the global warming treaty during a gathering of international negotiators in Buenos Aires charged with hammering out the many details left unresolved in Kyoto. Yet as this and future meetings drag on, it is clear that the free-swinging Senate debate predicted by Al Gore will not take place, at least during the Clinton presidency. "We don't expect major breakthroughs," said Undersecretary of State Stuart Eizenstat, who headed the U.S. delegation to Buenos Aires as he did in Kyoto. Above all, there is no foreseeable resolution of one of the protocol's thorniest issues: How to persuade developing countries such as China and India that they must commit to reducing emissions under the treaty as well. Only Argentina, a U.S. ally, and Kazakhstan have voluntarily agreed to accept the emissions limits that are to apply to the industrialized West.

Meanwhile, the newly published findings of two research teams have added more force to the argument that Earth is indeed warming. Temperatures taken from hundreds of bore holes drilled in Europe, Australia, North America, South Africa, Greenland, and Antarctica reveal that the average global temperature has increased by about 1.8° over the past five centuries. And about 80 percent of the warming has occurred since 1750, the start of the industrial age and the large-scale burning of fossil fuels. Half of that increase—or a little less than 1°F—has taken place since 1900. In the words of Henry Pollack, a professor at the University of Michigan and one of the team leaders, "The Earth seems to have developed a fever," the very conclusion reached by Svante Arrhenius more than a century ago.

Bibliography

The literature on climate change is vast, and I have included only those sources most directly related to this work. A list of Internet Web sites follows the chapter bibliographies.

CHAPTER 1. THE GUILLOTINE AND THE BELL JAR
Fourier, Joseph. "Memoire sur les témperatures du globe terrestre et des espaces planétaires." *Memoires de l'Académie Royale des Sciences*, 7 (1827), 569–604.

———. "Remarques générales sur la température du globe terrestre et des espaces planétaires." *Annales de chimie et de physique*, 27 (1824), 136–67.

Grattan-Guiness, I. *Joseph Fourier, 1768–1830: A Survey of His Life and Work*. Cambridge, Mass.: MIT Press, 1972.

Herivel, John. *Joseph Fourier: The Man and the Physicist*. Oxford: Clarendon Press, 1975.

CHAPTER 2. THE CRYPTIC MOTH
Bishop, J. A., and Laurence M. Cook. "Moths, Melanism, and Clean Air." *Scientific American*, 232 (January 1975), 90–98.

Bowler, Peter J. *The Mendelian Revolution*. Baltimore: Johns Hopkins University Press, 1989.

Clark, Ronald W. *The Survival of Charles Darwin: A Biography of a Man and an Idea*. New York: Random House, 1984.

Darwin, Charles. *On the Origin of Species by Means of Natural Selection*. New York: New American Library of World Literature, 1958.

Desmond, Adrian, and James Moore. *Darwin: The Life of a Tormented Evolutionist*. New York: Warner Books, 1991.

Edleston, R. S. *"Amphydasis betularia."* *Entomologist*, 2 (1864–65), 150.

Gaskell, Elizabeth Cleghorn. *Mary Barton*. In *The Works of Mrs. Gaskell with Introductions by A. W. Ward*, vol. 1. London: Smith, Elder, 1906.

Grant, B. S.; D. F. Owen; and C. A.Clarke. "Parallel Rise and Fall of

Melanic Peppered Moths in America and Britain." *Journal of Heredity*, 87 (1966), 351–57.

Kittlewell, Bernard. "Darwin's Missing Evidence." *Scientific American*, 200 (March 1959), 48–53.

———. *The Evolution of Melanism: The Study of a Recurring Necessity with Special Reference to Industrial Melanism in the Lepidoptera*. Oxford: Clarendon Press, 1973.

Mendel, Gregor. *Experiments in Plant Hybridisation*. Cambridge, Mass.: Harvard University Press, 1965.

Olby, Robert C. *Origins of Mendelism*. New York: Schocken Books, 1966.

CHAPTER 3. "ENDLESS AND AS NOTHING"

Bailey, Maurice E. *Coal and Other Rocks*. Pikeville, Ky.: Pikeville College Press, 1984.

Cohen, I. Bernard. *Benjamin Franklin: Scientist and Statesman*. New York: Scribner's, 1975.

Darwin, Charles. *The Voyage of the Beagle*. Garden City, N.Y.: Doubleday, 1962.

Franklin, Benjamin. *The Autobiography*. New York: Vintage Books, 1990.

———. *Observations on the Causes and Cure of Smoky Chimneys*. (self-published) Philadelphia, 1787.

Hutton, James. *Theory of the Earth with Proofs and Illustrations*. 2 vols. Edinburgh: Cadell, Junior and Davies, 1795.

Lyell, Charles. *Principles of Geology*. 3 vols. London: J. Murray, 1830–33.

Rudwick, M. J. S. *The Great Devonian Controversy: The Shaping of Scientific Knowledge Among Gentlemanly Specialists*. Chicago: University of Chicago Press, 1985.

CHAPTER 4. "QUEST FOR THE BLACK DIAMOND"

Burke, James. *Connections*. Boston: Little, Brown, 1978.

Burnett, John, ed. *Useful Toil: Autobiographies of Working People from the 1820s to the 1920s*. New York: Penguin, 1984.

Fraser, Grace Lovat. *Textiles in Britain*. London: Allen and Unwin, 1948.

Mann, Julia De Lacy. *The Cloth Industry in the West of England from 1640 to 1880*. Oxford: Clarendon Press, 1971.

Mantoux, Paul. *Industrial Revolution in the Eighteenth Century*. New York: Macmillan, 1961.

CHAPTER 5. CLEOPATRA'S NEEDLES

Bancroft, Robert M., and Francis J. Bancroft. *Tall Chimney Construction: A Practical Treatise on the Construction of Tall Chimney Shafts . . . in Brick, Stone, Iron, and Concrete*. Manchester: John Calvert, 1885.

Christie, William Wallace. *Chimney Design and Theory: A Book for Engineers and Architects.* New York: Van Nostrand, 1899.

Dickens, Charles. *Hard Times.* New York: Dutton, 1920.

Douet, James. *Going Up in Smoke: The History of the Industrial Chimney.* London: Victorian Society, 1989.

Ruskin, John. *The Two Paths: Being Lectures on Art and Its Applications to Decorations and Manufacture Delivered in 1858–9.* New York: Merrill and Baker, 1900.

Stevenson, Robert Louis. *An Inland Voyage.* In *The Works of Robert Louis Stevenson*, ed. Lloyd Osbourne, vol. 1. New York: Scribner's, 1921.

CHAPTER 6. VULCAN'S ANVIL

Fisher, Douglas Alan. *The Epic of Steel.* New York: Harper and Row, 1963.

Landes, David. *The Unbound Prometheus.* Cambridge, England: Cambridge University Press, 1969.

Vialls, Christine. *Coalbrookdale and the Iron Revolution.* Cambridge, England: Cambridge University Press, 1980.

CHAPTER 7. THE PHANTOM OF THE OPEN HEARTH

Carnegie, Andrew. *Autobiography of Andrew Carnegie.* Boston: Houghton Mifflin, 1920.

Goldin, Milton. "The Gospel of Andrew Carnegie." *History Today*, 38 (June 1988), 11–17.

Gould, Stephen Jay. "The Great Western and the Fighting Temeraire." *Natural History*, 104 (October 1995), 16–19, 62–65.

Kihlstedt, Folke T. "The Crystal Palace." *Scientific American*, 251 (October 1984), 132–43.

Petroski, Henry. "The Amazing Crystal Palace." *Technology Review*, 86 (July 1983), 18–28.

Shepherd, Jean. *Phantom of the Open Hearth.* Garden City, N.Y.: Doubleday, 1978.

Steward, Doug. "There Was Too Much Jonah in Brunel's Hapless Leviathan." *Smithsonian*, 25 (November 1994), 62–75.

Wall, Joseph Frazier. *Andrew Carnegie.* New York: Oxford University Press, 1970.

CHAPTER 8. "THE DYNAMO AND THE VIRGIN"

Adams, Henry. *The Education of Henry Adams: An Autobiography.* Boston: Houghton Mifflin, 1918.

Basalla, George. *The Evolution of Technology.* Cambridge, England: Cambridge University Press, 1988.

Csere, Csaba. "Ten Best Engineers." *Car and Driver*, 31 (January 1986), 48–51.

Derry, Thomas K., and Trevor I. Williams. *A Short History of Technology.* Oxford: Oxford University Press, 1961.

Hohman, Paul Elmo. *The American Whaleman: A Study of the Life and Labor in the Whaling Industry.* New York: Longmans, Green, 1928.

Lord Montagu of Beaulieu. "The Early Days of Motoring." *History Today,* 36 (October 1986), 43–49.

Nevins, Allan. *Study in Power: John D. Rockefeller, Industrialist and Philanthropist.* New York: Scribner's, 1953.

Schultz, Mort. "Engines: A Century of Progress." *Popular Mechanics,* 162 (January 1985), 95–97, 120–22.

Vernon, Bill. "Probing Earth's History: Drake Well . . . Gave Birth to an Industry That Changed World Cultures." *Earth Science,* 41 (Spring 1988), 12–15.

CHAPTER 9. NATIVE SON

Arrhenius, Svante. "Electrolytic Dissociation." *Journal of the American Chemical Society,* 34 (April 1912), 353–65.

———. "On the Influence of Carbonic Acid in the Air upon the Temperature of the Ground." *The London, Edinburgh, and Dublin Philosophical Magazine and Journal of Science,* 5th ser. (April 1896), 237–76.

———. *Worlds in the Making: The Evolution of the Universe.* New York: Harper and Brothers, 1908.

Crawford, Elisabeth. *Arrhenius: From Ionic Theory to the Greenhouse Effect.* Canton, Mass.: Science History Publications, 1996.

Furneaux, Rupert. *Krakatoa.* Englewood Cliffs, N.J.: Prentice-Hall, 1964.

Harrow, Benjamin. *Eminent Chemists of Our Time.* New York: Van Nostrand, 1920.

Kauffman, George B. "Svante August Arrhenius: Swedish Pioneer in Physical Chemistry." *Journal of Chemical Education,* 65 (May 1988), 437–38.

Kellogg, William W. "Mankind's Impact on Climate: The Evolution of an Awareness." *Climate Change,* 10 (1987), 113–36.

Langley, S. P. "Researches on Solar Heat and Its Absorption by the Earth's Atmosphere: A Report of the Mount Whitney Expedition." *Professional Papers of the Signal Service,* 15 (1884), 1–242.

Simkin, Tom, and Richard S. Fiske. *Krakatau 1883: The Volcanic Eruption and Its Effects.* Washington, D.C.: Smithsonian Institution Press, 1983.

Thornton, Ian. *Kratatau: The Destruction and Reassembly of an Island Ecosystem.* Cambridge, Mass.: Harvard University Press, 1995.

Tyndall, John. "On the Absorption and Radiation of Heat by Gases and Vapours, and on the Physical Connexion of Radiation, Absorption, and Conduction." *The London, Edinburgh, and Dublin Philosophical Magazine and Journal of Science,* 4th ser. (September 1861), 169–94.

Uppenbrink, Julia. "Arrhenius and Global Warming." *Science,* 272 (May 24, 1996), 1122.

Walker, James. "Arrhenius Memorial Lecture." *Journal of the Chemical Society*, 1, pt. 1 (1928), 1380–1401.

CHAPTER 10. "NEVER A MAN"

Brent, Peter. *The Viking Saga.* New York: Putnams, 1975.

Brody, J. J. "The Chaco Phenemonon." *Archaeology News*, 36 (July–August, 1983), 57–61.

Ferguson, William M. *The Anasazi of Mesa Verde and the Four Corners.* Niwot, Colo.: University Press of Colorado, 1996.

Gribbon, John, and Mary Gribbon. "Climate and History: The Westvikings' Saga." *New Scientist*, 125 (January 20, 1990), 52–55.

Hass, Jonathan, and Winifred Creamer. "Stress and Warfare Among the Kayenta Anasazi of the Thirteenth Century A.D." *Fieldiana: Anthropology*, n.s., 21 (1993), 1–211.

Johnson, George. "Social Strife May Have Exiled Ancient Indians." *New York Times*, August 20, 1996, sec. C, pp. 1, 6.

Jones, Gwyn. *A History of the Vikings.* London: Oxford University Press, 1968.

Larson, Daniel O.; Hector Neff; Donald A. Grabill; Joel Michaelsen; and Elizabeth Ambos. "Risk, Climate Variability, and the Study of Southwestern Prehistory: An Evolutionary Perspective." *American Antiquity*, 61 (April 1996), 217–41.

Lekson, Stephen. "Tracking the Movements of an Ancient People." *Archaeology*, 48 (September–October, 1995), 56–57.

Magnussen, Magnus. *Vikings!* New York: Dutton, 1980.

McDonald, Kim A. "A Lost Settlement in Greenland and Climate Change." *Chronicle of Higher Education*, November 15, 1966, sec. A, pp. 15, 18.

———. "Preserving a Priceless Library of Ice." *Chronicle of Higher Education*, August 2, 1996, sec. A, pp. 7, 11.

———. "Unearthing Earth's Ancient Atmosphere Beneath Two Miles of Greenland Ice." *Chronicle of Higher Education*, August 2, 1996, sec. A, pp. 6, 11.

Monastersky, R. "Viking Teeth Recount Sad Greenland Tale." *Science News*, 146 (November 12, 1994), 310.

Parfit, Michael. "The Dust Bowl." *Smithsonian*, 20 (June 1989), 44–54.

Peterson, Kenneth Lee. "A Warm and Wet Little Climatic Optimum and a Cold and Dry Little Ice Age in the Southern Rocky Mountains, U.S.A." *Climatic Change*, 26 (March 1994), 243–69.

Pringle, Heather. "Death in Norse Greenland." *Science*, 275 (February 14, 1997), 924–26.

Roberts, David. "The Old Ones of the Southwest." *National Geographic*, 189 (April 1996), 86–109.

Schlanger, Sarah H. "Patterns of Population Movement and Long-Term Population Growth in Southwestern Colorado." *American Antiquity*, 53 (October 1988), 773–93.

Trewartha, Glenn T. *An Introduction to Weather and Climate*. New York: McGraw-Hill, 1937.

CHAPTER 11. THRESHOLD

Arrhenius, Svante. *Chemistry in Modern Life*. New York: Van Nostrand, 1926.

Callendar, G. S. "The Artificial Production of Carbon Dioxide and Its Influence on Temperature." *Quarterly Journal of the Royal Meteorological Society*, 64 (1938), 223–40.

———. "Can Carbon Dioxide Influence Climate?" *Weather*, 4 (1949), 310–14.

———. "Variations of the Amount of Carbon Dioxide in Different Air Currents." *Quarterly Journal of the Royal Meteorological Society*, 66 (1940), 395–400.

Christianson, Gale E. *Edwin Hubble: Mariner of the Nebulae*. New York: Farrar Straus and Giroux, 1995.

Mintz, Morton. "GM Accused of Helping Destroy Electric Rail Transit in 45 Cities." *Los Angeles Times*, February 25, 1974, sec. I, pp. 1, 10.

Page, Clint. "Return Trip for the Trolley." *Nation's Cities*, 15 (October 1977), 4–7, 16.

CHAPTER 12. A TAP ON THE SHOULDER

Brimblecombe, Peter. *The Big Smoke: A History of Air Pollution in London Since Medieval Times*. London: Methuen, 1987.

Callendar, G. S. "On the Amount of Carbon Dioxide in the Atmosphere." *Tellus*, 10 (1958), 243–48.

Haynes, William. "Thomas Midgley, Jr." In *Great Chemists*, ed. Eduard Farber, 1588–97. New York: Interscience Publishers, 1961.

Jones, M. D. H., and A. Henderson-Sellers. "History of the Greenhouse Effect." *Progress in Physical Geography*, 14 (1990), 1–18.

Keeling, Charles D. "A Chemist Thinks About the Future." *Archives of Environmental Health*, 20 (1970), 764–77.

Revelle, Roger, and Hans S. Suess. "Carbon Dioxide Exchanges Between Atmosphere and Ocean and the Question of an Increase of Atmospheric CO_2 During the Past Decades." *Tellus*, 9 (1957), 18–27.

The Times of London, December 4 to 10, 1952.

Weiner, Jonathan. *The Next One Hundred Years: Shaping the Fate of Our Living Earth*. New York: Bantam Books, 1990.

CHAPTER 13. PENDULUM

Abbey, Edward. *The Monkey Wrench Gang*. Philadelphia and New York: Lippincott, 1975.

Alexander, George. "Colder Winters Ahead?" *Popular Science*, 211 (October 1977), 100–102.

Broecker, Wallace S. "Climatic Change: Are We on the Brink of a Pronounced Global Warming?" *Science*, 189 (August 8, 1975), 460–63.

Bryson, Reid A., and Brian M. Goodman. "Volcanic Activity and Climatic Changes." *Science*, 207 (March 7, 1980), 1041–44.

Clemens, Steven C., and Ralf Tiedemann. "Eccentricity Forcing of Pliocene–Early Pleistocene Climate Revealed in a Marine Oxygen-Isotope Record." *Nature*, 385 (February 27, 1997), 801–4.

Ehrlich, Paul. *The Population Bomb*. New York: Ballantine Books, 1968.

Gwynne, Peter. "The Cooling World," *Newsweek*, (April 28, 1975), 64.

Hays, J. D.; John Imbrie; and N. S. Shakelton. "Variations in the Earth's Orbit: Pacemaker of the Ice Ages." *Science*, 194 (December 10, 1974), 1121–32.

Kerr, Richard A. "Ice Rhythms: Core Reveals a Plethora of Climate Cycles." *Science*, 274 (October 25, 1996), 499–500.

Lamb, H.H. *Climate, History, and the Modern World*. London: Methuen, 1982.

Levinson, Thomas. *Ice Time: Climate, Science, and Life on Earth*. New York: Harper and Row, 1989.

Matthews, Samuel. "What's Happening to Our Climate?" *National Geographic*, 150 (November 1976), 576–615.

Molina, Mario J., and F. S. Rowland."Stratospheric Sink for Chlorofluoromethanes: Chlorine Atom–Catalysed Destruction of Ozone." *Nature*, 249 (June 28, 1974), 810–12.

Muller, Richard A., and Gordon J. MacDonald. "Glacial Cycles and Astronomical Forcing." *Science*, 277 (July 11, 1997), 215–18.

Penvenne, Laura Jean. "The Bolide and the Biosphere." *American Scientist*, 85 (January–February 1997), 25.

Ramanathan, V. "Greenhouse Effect Due to Chlorofluorocarbons: Climatic Implications." *Science*, 190 (October 1975), 50–52.

Rifkin, Jeremy. *Entropy: A New World View*. New York: Viking Press, 1980.

Salinger, M. J., and J. M.Gunn. "Recent Climatic Warming Around New Zealand." *Nature*, 256 (July 31, 1975), 396–98.

Somerville, Richard C. J. *The Forgiving Air: Understanding Environmental Change*. Berkeley and Los Angeles: University of California Press, 1996.

Weiner, Jonathan. *The Next One Hundred Years: Shaping the Fate of Our Living Earth*.

Wolff, Anthony. "Are We Headed for a New Ice Age?" *Current*, 183 (May–June, 1976), 56–60.

Chapter 14. A Death in the Amazon

Abrahamson, Dean Edwin. *The Challenge of Global Warming*. Washington, D.C.: Island Press, 1989.

Carlson, Shawn. "Death in the Rain Forests." *Humanist*, 52 (March–April, 1992), 35–37.

Dickens, Charles. *American Notes and Pictures from Italy*. London: Oxford University Press, 1957.

Dotto, Lydia, and Harold Schiff. *The Ozone War*. Garden City, N.Y.: Doubleday, 1978.

Farman, J. C.; B. G. Gardiner; and J. D. Shanklin. "Large Losses of Total Ozone in Antarctica Reveal Seasonal CIO$_x$/NO$_x$ Interaction." *Nature*, 315 (May 16, 1985), 207–10.

Firor, John. *The Changing Atmosphere*. New Haven: Yale University Press, 1990.

Frazier, Ian. *Family*. New York: Farrar Straus and Giroux, 1994.

"Galápagos." *New Yorker* (April 13, 1998), 82.

Gribbon, John, ed. *The Breathing Planet*. Oxford: Basil Blackwell, 1986.

Hadfield, Peter. "Forest Watchdog Fails to Show Its Teeth." *New Scientist*, 132 (December 14, 1991), 12.

"It's Our Forest to Burn If We Want To." *Economist*, 310 (March 11, 1989), 42–43.

Killian, Linda. "Dead Effort." *Forbes*, 147 (June 24, 1991), 96.

Markham, Adam. *A Brief History of Pollution*. New York: St. Martin's Press, 1994.

Marsh, George P. *The Earth as Modified by Human Action: A Last Revision of "Man and Nature."* New York: Scribner's, 1885.

"The Month the Amazon Burns." *Economist*, 312 (September 9, 1989), 15–16.

Nance, John J. *What Goes Up: The Global Assault on Our Atmosphere*. New York: Morrow, 1991.

Oppenheimer, Michael, and Robert H. Boyle. *Dead Heat: The Race Against the Greenhouse Effect*. New York: Basic Books, 1990.

"Ozone: The Crisis That Wasn't." *Science Digest*, 92 (August 1984), 30.

Park, Chris C. *Acid Rain: Rhetoric and Reality*. London: Methuen, 1987.

Pearce, Fred. "First Aid for the Amazon." *New Scientist*, 133 (March 28, 1992), 42–45.

———. "Hit and Run in Sarawak. *New Scientist*, 126 (May 12, 1990), 46–49.

Ponting, Clive. *A Green History of the World*. New York: St. Martin's Press, 1991.

Revkin, Andrew C. "Murder in the Amazon." *Discover*, 11 (January 1990), 30–32.

Sanford, Robert L.; Juan Saldarriaga; Kathleen Clark; Christopher Uhl; and Raphael Herrera. "Amazon Rain-Forest Fires." *Science*, 227 (January 4, 1985), 53–55.

Sattaur, Omar. "Last Chance for the Rainforest Plan?" *New Scientist*, 129 (March 2, 1991), 20–21.

Schemo, Diana Jean. "Brazil Says Amazon Burning Tripled in Recent Years." *New York Times*, January 27, 1988, sec. A, p. 3.

Scott, Margaret. "The Disappearing Forests." *Far Eastern Economic Review*, 143 (January 1989), 34–38.

Sioli, Harald. "The Effects of Deforestation in Amazonia." *Geographical Journal*, 151 (July 1985), 197–203.

Smith, Robert Angus. *Air and Rain: The Beginnings of a Chemical Climatology*. London: Longmans Green, 1872.

———. "On the Air and Rain of Manchester." *Memoirs and Proceedings of the Manchester Literary and Philosophical Society*, 2 (1852), 207–17.

Turco, Richard P. *Earth Under Siege: From Air Pollution to Global Change*. Oxford: Oxford University Press, 1997.

Wallace, Alfred Russel. *The Malay Archipelago*. 1869. Reprint, New York: Dover, 1962.

———. *Travels on the Amazon and Rio Negro*. London: Ward, Lock, 1889.

Wilson, A. T. "Pioneer Agriculture Explosion and CO_2 Levels in the Atmosphere." *Nature*, 273 (May 4, 1978), 40–41.

CHAPTER 15. THE CLIMATIC FLYWHEEL

Balling, Robert C., Jr. *The Heated Debate: Greenhouse Predictions Versus Climate Reality*. San Francisco: Pacific Research Institute for Public Policy, 1992.

Beardsley, Tim. "Getting Warmer? This Has (So Far) Been the Warmest Decade in 127 Years." *Scientific American*, 259 (July 1988), 32.

———. "Winds of Change: International Talks Address Human Effects on Climate." *Scientific American*, 259 (September 1988), 18–19.

Fisher, Arthur. "Global Warming: Playing Dice with Earth's Climate." *Popular Science*, August 1989, 51–58.

Heim, Richard R., Jr. "About that Drought." *Weatherwise*, 41 (October, 1988), 266–70.

Horgan, John. "Pinning Down Clouds: Scientists Ponder the Role of Clouds in Climatic Change." *Scientific American*, 260 (May 1989), 22–24.

Jones, P. D.; T. M. L. Wigley; and P. B. Wright. "Global Temperature Variations Between 1861 and 1984." *Nature*, 322 (July 31, 1986), 430–34.

Kerr, Richard A. "Greenhouse Forecasting Still Cloudy." *Science*, 276 (May 16, 1997), 1040–42.

———. "Greenhouse Skeptic Out in the Cold." *Science*, 246 (December 1, 1989), 1118–19.

———. "Hansen vs. the World on the Greenhouse Threat." *Science*, 244 (June 2, 1989), 1041–43.

McKibben, Bill. *The End of Nature*. New York: Random House, 1989.

Nance, John J. *What Goes Up: The Global Assault on Our Atmosphere*.

Roberts, Leslie. "Global Warming: Blaming the Sun." *Science*, 246 (November 24, 1989), 992–93.

Rossow, William B. "Who Knows Where the Clouds Go?" *The Sciences*, 31 (May–June 1991), 36–41.

Schneider, Stephen H. *Global Warming: Are We Entering the Greenhouse Century?* San Francisco: Sierra Club Books, 1989.

Shabecoff, Philip. "Dozens of Nations Reach Agreement to Protect Ozone." *New York Times*, September 17, 1987, sec. A, p. 1.

Sjvøld, Thorstein. "Frost and Found." *Natural History*, 102 (April 1993), 60–63.

Somerville, Richard C. J. *The Forgiving Air: Understanding Environmental Change.*

Stevens, William K. "At Hot Center of Debate on Global Warming." *New York Times*, August 26, 1996, sec. C, pp. 1, 10.

Stouffer, R. J.; S. Manabe; and K. Y. Vinnikov. "Model Assessment of the Role of Natural Variability in Recent Global Warming." *Nature*, 367 (February 17, 1994), 634–36.

Trenberth, Kevin E. "The Use and Abuse of Climate Models." *Nature*, 386 (March 13, 1997), 131–34.

CHAPTER 16. CASSANDRA'S LISTENERS

Anderson, Ian. "Global Warming Rings True." *Science*, 131 (September 1991), 23.

Brower, B. "CO_2 Rises Stretch Tree-ring Sizes." *Science News*, 126 (September 8, 1984), 156.

Brown, Barbara E., and John C. Ogden. "Coral Bleaching." *Scientific American*, 268 (January 1993), 64–70.

Carlowicz, Michael. "Did Water Vapor Drive Climate Cooling?" *EOS*, 77 (August 13, 1996), 321–22.

Dacy, J. W. H.; B. G. Drake; and M. J. Klug. "Stimulation of Methane Emission by Carbon Dioxide Enrichment of Marsh Vegetation." *Nature*, 370 (July 7, 1994), 47–49.

Darwin, Charles. *The Voyage of the Beagle.*

"Global Warming: Beyond Termites." *Science News*, 145 (June 25, 1994), 410.

Hileman, Bette. "Role of Methane in Global Warming Continues to Perplex Scientists." *Chemical and Engineering News*, 70 (February 10, 1992), 26–28.

Jayaraman, K. S. "Skinny Cows No Threat." *Nature*, 353 (October, 24, 1991), 685.

Johannessen, Ola M.; Einar Bjørgo; and Martin W. Miles. "The Arctic's Shrinking Sea Ice." *Nature*, 376 (July 13, 1995), 126–27.

———. "Global Warming and the Arctic." *Science*, 271 (January 12, 1996), 129.

"Limiting Rice's Role in Global Warming." *Science News*, 144 (July 10, 1993), 30.

Mitchell, John F. B. "The 'Greenhouse' Effect and Climate Change." *Reviews of Geophysics*, 27 (February 1, 1989), 115–39.

Raloff, J. "Butterfly Displaced by Climate Change?" *Science News*, 150 (August 31, 1996), 135.

Sagan, Carl. "Croesus and Cassandra: Policy Response to Global Climate Warming." *American Journal of Physics*, 58 (August 1990), 721–30.

Stevens, William K. "Western Butterfly Shifting North as Global Climate Warms." *New York Times*, September 2, 1996, sec. C, p. 4.

Stone, Richard. "Rustic Site Draws a Crowd to Monitor Global Warming." *Science*, 266 (October 1994), 360–61.

Weisburd, S. "Greenhouse Gases En Masse Rival CO_2." *Science News*, 127 (May 18, 1985), 308.

Williams, Jack. "Cold to the Core While Studying Greenland's Ice." *New York Times*, April 28, 1988, sec. D, p. 8.

Zwally, H. J. "Breakup of Antarctic Ice." *Nature*, 350 (March 28, 1991), 274.

CHAPTER 17. SIGNS AND PORTENTS

"An Act of God." *Economist*, 344 (July 19, 1997), 69–71.

"Africa with No Rain, or too Much." *Economist*, 345 (December 13, 1997), 39.

Cavides, Cesar. "El Niño 1982–3." *Geographical Review*, 74 (July 1984), 267–90.

DePalma, Anthony. "Freeze Grips Canada, Already Iced and Powerless." *New York Times*, January 13, 1998, sec. A, p. 1.

"Early Winter Storm Kills 3,000 Calves on Ranch in New Mexico." *New York Times*, December 26, 1997, sec. A, p. 26.

Green, Emily. "Food." *New Statesman*, 127 (February 20, 1998), 39.

Henson, Bob. "Global Temperatures from Space and the Surface: Why the Discrepancy?" *UCAR Quarterly*, 21 (Winter–Spring 1997), 1, 8–9.

Kerr, Richard A. "Global Temperature Hits Record Again." *Science*, 251 (January 18, 1991), 274.

McKinley, James C., Jr. "Rain Is a New Agony for Somalia as Villages Are Suddenly Islands." *New York Times*, November 19, 1997, sec. A, p. 1.

Meehl, Gerald, and Warren M. Washington. "El Niño-like Climate Change in a Model with Increased Atmospheric CO_2 Concentrations." *Nature*, 382 (July 4, 1996), 55–60.

Monastersky, R. "El Niño Shifts Earth's Momentum." *Science News*, 153 (January 17, 1998), 45.

———. "Global Temperatures Spark Hot Debate." *Science News*, 151 (March 15, 1997), 156.

———. "1995 Captures Record as Warmest Year Yet." *Science News*, 149 (January 13, 1996), 23.

Mydans, Seth. "Southeast Asia Chokes on Indonesia's Forest Fires." *New York Times*, September 25, 1997, sec. A, p. 1.

Pain, Stephanie. "Darwin's Paradise Awash." *New Scientist*, 157 (January 10, 1998), 4.

Schumacher, Edward. "Floods and Droughts Sweep Across South America." *New York Times,* June 12, 1983, sec. A, p. 1.

Sehgal, R. "Dengue Fever and El Niño." *Lancet*, 349 (March 8, 1997), 729–30.

Travis, J. "The Loitering El Niño: Greenhouse Guest?" *Science News*, 149 (January 27, 1996), 54.

Webster, Ferris. "Studying El Niño on a Global Scale." *Oceanus*, 27 (Summer 1984), 58–62.

CHAPTER 18. SCENARIOS

"Adapting to Global Warming Will Be Easy, Says Archaeologist." *New Scientist*, 125 (February 24, 1990), 24.

Bernard, Harold W., Jr. *Global Warming Unchecked: Signs to Watch For.* Bloomington: Indiana University Press, 1993.

Brown, Paul. *Global Warming: Can Civilization Survive?* London: Blanford, 1966.

Cushman, John H., Jr. "Texans Coping With Smoke Cloud From Fires in Mexico." *New York Times,* May 18, 1998, sec A, p. 10.

Dillon, Sam. "A Long Fire Season Torments Mexico." *New York Times,* May 16, 1998, sec. A, p. 4.

Dobson, Andrew. "Withering Heats." *Natural History*, 101 (September 1992), 2–8.

Hansen, James; Makiko Sato; Reto Ruedy; Andrew Lacis; and Jay Glascoe. "Global Climate Data and Models: A Reconciliation." *Science,* 128 (August 14, 1998), 930–32.

Horgan, J. T. "Greenhouse America: A Global Warming May Destroy U.S. Forests and Wetlands." *Scientific American*, 260 (January 1989), 20–21.

Houghton, J. T.; G. J. Genkins; and J. J.Ephraums, eds. *Climate Change: The IPCC Scientific Assessment.* Cambridge, England: Cambridge University Press, 1990.

Houghton, J. T.; L. G. Meira Filho; J. Bruce; Joesung Lee; B. A. Callender; E. Haites; N. Harris; and K. Maskell, eds. *Climate Change 1994: Radiative Forcing of Climate Change and an Evaluation of the IPCC IS92 Emission Scenarios.* Cambridge, England: Cambridge University Press, 1995.

Kristof, Nicholas D. "Island Nations Fear Sea Could Swamp Them." *New York Times*, December 1, 1997, sec. F, p. 9.

Moore, Thomas Gale. "Why Global Warming Would Be Good For You." *Public Interest*, 118 (Winter 1995), 83–99.

Singer, S. Fred. *Hot Talk, Cold Science: Global Warming's Unfinished Debate.* Oakland, Calif.: Independent Institute, 1997.

Stevens, William K. "As Debate Persists, New Study Confirms Atmospheric Warming." *New York Times,* August 13, 1998, sec. A, p. 12.

"Weather," *USA Today,* October 6, 1997, sec. A, p. 5.

Wentz, Frank J., and Matthias Schnabel. "Effects of Orbital Decay on Satellite-Derived Lower-Tropospheric Temperature Trends." *Nature,* 394 (August 13, 1998), 661–63.

Wilber, John Noble. "Showing Why a Rain Forest Matters." *New York Times,* May 29, 1998, sec. B, p. 31.

CHAPTER 19. KYOTO

"The Apocalypse and Al Gore." ABC News Saturday Night, Peter Jennings Reporting: Landover, Md.: Federal Document Clearing House, April 11, 1998.

Clinton, William J., and Albert Gore, Jr. *The Climate Change Action Plan.* Washington, D.C.: Executive Office of the President, 1993.

Cushman, John H., Jr. "Pressure Points in Global Warming." *New York Times,* December 7, 1997, sec. A, p. 23.

Gore, Al. *Earth in the Balance: Ecology and the Human Spirit.* Boston: Houghton Mifflin, 1992.

Kurzman, Dan. *Day of the Bomb: Countdown to Hiroshima.* New York: McGraw-Hill, 1986.

"A Preview to the Kyoto Conference: Global Warming." *New York Times,* December 1, 1997, sec. F, p. 1.

Watson, Traci. "Global Warming Deal Required a World of Give-and-Take." *USA Today,* December 11, 1997, sec. A, p. 6.

CODA

Bradsher, Keith. "Light Trucks Have Passed Cars on the Retail Sales Road." *New York Times,* December 4, 1997, sec.D, p. 4.

———. "Plan Allows Big Vehicles to Skirt Rules on Pollutants." *New York Times,* December 19, 1997, sec. A, p. 17.

Brooke, James. "Redwoods Still Inspire Sturdiest of Defenders." *New York Times,* March 28, 1998, sec. A, p. 6.

Cushman, John H., Jr. "Brazil to Set Aside Vast Tract in the Amazon for Conservation." *New York Times,* April 30, 1998, sec. A, p. 5.

———. "Industrial Group Plans to Battle Climate Treaty." *New York Times,* April 26, 1998, sec. A, pp. 1, 17.

———. "Scientists Are Turning to Trees to Repair the Greenhouse." *New York Times,* March 3, 1998, sec. B, p. 15.

Eldridge, Earle. "Big 3: Global-Warming Pact Poses Threat." *USA Today,* December 12, 1997, sec. B, p. 7.

Kasindorf, Martin. "Sibling Rivalry Is Brewing: Exit El Niño, Enter La Niña." *USA Today,* June 11, 1998, sec. A, p. 14.

Park, Robert L. "Scientists and Their Political Passions." *New York Times,* May 2, 1998, sec. A, p. 23.

Pollack, Henry N.; Shaopeng Huang; and Po-Yu Shen. "Climate Change Record in Subsurface Temperatures: A Global Perspective." *Science,* 282 (October 9, 1998), 279–81.

Schemo, Diana Jean. "Brazil Says Amazon Burning Tripled in Recent Years."

Stevens, William K. "Global Temperature at a High for the First Five Months of 1998." *New York Times,* June 8, 1998, sec. A, pp. 1, 4.

————. "Science Academy Disputes Attack on Global Warming." *New York Times,* April 22, 1998, sec. A, p. 20.

Watson, Traci. "July Breaks Worldwide Temperature Record." *USA Today,* August 11, 1998, sec. A, p. 3.

————. "New Investment for Corporations: Protecting Trees." *USA Today,* April 14, 1998, sec. A, p. 5.

WEB SITES

Energy Efficiency and Renewable Energy Resources Network: www.eren.doe

Environmental Defense Fund: www.crl.com

George C. Marshall Institute: www.marshall.org

Global Change Master Directory: http://gemd.gsfac.nasa.gov

Intergovernmental Panel on Climate Change: www.ipcc.ch

Sierra Club: www.sierraclub.org

Union of Concerned Scientists: www.ucsua.org

United States Department of Energy: www.doe.gov

United States Environmental Protection Agency: www.epa.gov

Washington Post: http://search.washingtonpost.com

World Climate Report: www.nhes.com

Acknowledgments

Libraries constitute the lifeblood of scholarship, and I must begin by thanking the faculty and staff of the Cunningham Memorial Library at Indiana State University for their invaluable assistance. I am especially grateful to Dara Middleton, formerly of the Interlibrary Loan department, who assisted in tracking down a number of my more obscure sources. Graduate students Bruce Dickerson and George Stachokas proved themselves excellent bibliographers and research assistants. Librarian Lynn Joshi of the Hagley Museum and Library provided some excellent sources in the history of technology, including the loan of some rare books on tall chimney construction.

Dr. Henri Dee Grissino-Mayer of Valdosta State University helped me to better understand the more technical aspects of tree-ring dating. Anthropologist Dr. Winifred Creamer of Northern Illinois University graciously answered a number of questions about Anasazi culture in the Four Corners region. At my own university, climatologist Dr. John Oliver must at times have thought of himself as my coauthor, so many were the questions he answered with utmost patience and accuracy. The same can be said for my colleagues Dr. John Allen, Dr. George Baaken, Dr. Greg Bierly, Dr. William Dando, and Dr. Basil Gomez, outstanding scholars all. Needless to say, the responsibility for any errors, however unintentional, is entirely my own.

A special thank-you goes to Nancy Miller of Walker and Company for supplying my agent with four pages of photocopied material on the Nobel laureate Svante Arrhenius and asking, "I wonder if Gale Christianson might be interested in doing something on this?" Publisher George Gibson was highly supportive from the outset, and I thank him for his faith in both the author and his subject. My editor, Jacqueline Johnson, is a perceptive and discerning spirit endowed with the patience of Job. Nor will I forget the many kindnesses of the others employed by Walker: Christopher Carey, Beth Caspar, Cassie Dendurent, Vicki Haire, Ivy Hamlin, Judi Ellen Kloos, Eileen Laverty Pagan, Krystyna Skalski, and Marlene Tungseth. Thanks also to my literary agent, Michael Congdon, with whom I have traveled so many miles. To Rhonda Packer, my wife and partner in scholarship, as in so much else, I have no words to equal my gratitude or affection. And finally, thanks to Manny Bear, whose long hours of silent companionship while I was at the computer are mine to cherish. After five books and ten years, I know you had to go. Bless you for waiting until the final chapter was almost done. Rest in peace my great heart, my bullyboy.

Index

Abbey, Edward, 165
Académie des Sciences, 3, 10–11
Académie Française, 12
Acid rain, 176–81, 190, 215
Adams, Henry, xiii, 101–2
Adams, John, 267
Adirondacks, 179
"Adventure of the Bruce–Partington Plans, The" (Conan Doyle), 149
Africa
  El Niño in, 226–27
  ozone levels, 247
Agriculture, 243, 251–52
Air and Rain (Smith), 177
Albatrosses, 225
Albedo, 164, 202, 203, 252
Albert, Prince, 75
Alexander the Great, 94
Algae, 225
Alliance of Small Island States, 256
Amazon, xiv, 188, 189, 190, 191, 195, 272
Amazon Pact, 189–90
Amazon River basin, 184
American Electric Power, 271
American Indians, 94, 117–72, 132
American Medical Association, 146
American Notes (Dickens), 181–82
Amundsen, Roald, 173
"Analytical Theory of Heat" (Fourier), 11
Anasazi ("Ancient Ones"), xiii, 117–22, 124, 125, 129, 132, 161, 249
Andrews, Samuel, 97
Annales de chimie et de physique, 12
Antarctica, 172–76, 203, 261
Arctic, 217, 218
Arctic Ocean, 92
Arkwright, Richard, xiii, 46–48, 49, 50, 52
Arrhenius, Svante August, xiii, xiv, 105–10, 111, 112, 113–15, 142, 143, 144, 155, 160, 251, 277
Arrhenius, Svante Gustave, 107
"Artificial Production of Carbon Dioxide

and Its Influence on Temperature, The" (Callendar), 141
Asvaldsson, Thorvald, 122
Atlantic cable, 83
Atmosphere, xii, 11, 200
  angular momentum, 227
  CO_2 in, xiii, xiv, 30–31, 33, 111, 141–42, 189, 202, 204, 215, 224, 231, 249–50
  cooling of, 120, 160
  heat–absorbing gases in, 109–10, 114, 132
  thickness of, 211
Atmospheric pollution, 20, 21, 23
Atmospheric warming, xi, 115
  benign influence of, 142
Austin, Mary, 120
Australia, 183, 228, 252, 257
Automobile, 101, 102, 137, 138–140, 143, 177, 274

Bacon, Sir Francis, 269
Balling, Robert, 259
Bancroft, Francis, 55
Bancroft, Robert M., 55
Bank Works of London, 63–64
Bárdarson, Ivar, 122
Barnett, Tim P., 224
Becker, Dan, 271
Bell jar hypothesis, 11–12, 113
Benedick, Richard E., 195
Benz, Karl, 101
Bering Strait, 92, 93
"Berlin Mandate," 264
Bessemer, Henry, xiii, 83–87, 89, 168
Bible, 24, 25–26
Biodiversity, treaties on, 262
Biomass burning, 247–48
Bjørgo, Einar, 218
Black Forest, 180
Blackwelder, Brent, 272
Blair, Tony, 265

Bluetongue, 226–27
Bolin, Bert, 263
Boltzmann, Ludwig, 109
Borja, Rodrigo, 189–90
Boulton, Matthew, 53, 71
Brazil, 15, 188, 190–91, 247, 272
Brent, Peter, 128
British Antarctic Survey, 173–74, 195
British Association for the Advancement
  of Science, 83
Broecker, Wallace S., 171, 199–200, 205,
  221
Bronze, 67
Brooks, C. E. P., 163–64
Brooks Range, 216
Brown, Harrison, 151–52
Brueghel, Pieter, 161
Brunel, Sir Isambard Kingdom, xiii,
  80–83
Brunel, Sir Marc Isambard, 80
Bumpers, Dale, 196
Burke, Edmund, 40–41
Burroughs, John, 198
Bush, George, 200, 261, 262
Butterflies, xiv, 13, 211–13, 215
Byrd, Richard, 156
Byron, Lord, 148

Callendar, George S., xiii, 141–42, 145,
  148, 151, 153–54, 155, 159, 167–68
Calley, John, 44
Capitalists, xiii, 48, 247
Carbon, 29–30, 31, 247
  sequestration of, 271–72
Carbon cycle, 30–31, 32, 157, 181
Carbon dioxide (CO_2), xiii, 30–31, 33,
  110, 149, 170, 171, 219
  in atmosphere, xiii, xiv, 30–31, 33, 111,
    141–42, 189, 202, 204, 215, 224, 231,
    249–50
  balance of, in atmosphere/oceans,
    152–53, 154, 155–57
  deforestation and, 189
  sensitivity of clouds to, 207
  sources of, 111
Carbon dioxide emissions, 144, 273
  reduction in, 256, 266
Carbon dioxide levels, 182–83
  rising, xiv, 148, 160, 162, 166–68
  scenarios regarding, 235, 236–37, 243,
    251–52
Carbon sink
  oceans as, 141–42, 154, 168, 189, 250
Carbonic acid
  effect on temperature, 113–15
Carboniferous period, 28, 31
Cardoso, Fernando Henrique, 272
Carlyle, Thomas, 98
Carnegie, Andrew, xiii, 87–91, 98, 168, 274

Carnegie, William, 88
Carnegie Steel Company, 90
Cartwright, Edmund, 49–51
Cast iron, 67–68, 77
Catalytic converter, 145
Catalytic cycle, 169
Catastrophism, 24–25, 164–65, 237
Cenozoic, 28, 31
Cereal production, 242–43
Chaco Phenomenon, 118–19
Challenge of Man's Future, The (Brown),
  152
Chancel, Gustav, 33
Charcoal, 42, 68–69, 70
Charles I, 42
Chemistry in Modern Life (Arrhenius),
  142
Chico Mendes Reserve, 191
Chicxulub crater, 31–32
Chimney Design and Theory (Christie),
  65
Chimneys, 34–35, 55–56, 149
  see also "Tall chimneys"
China, 167, 190, 264, 267, 275, 277
Chlorine, 169, 175
Chlorofluorocarbon industry, 174, 194,
  195
Chlorofluorocarbons (CFCs), xiv,
  146–47, 168–70, 174, 194–96, 261
Cholera, 226
Christie, William Wallace, 65
Civil disobedience, 165, 198–99
Civilizations
  climate change and, 129, 161, 162, 249
Clapp, Philip E., 267
Clean Air Act of 1956 (England), 151
Cleopatra's Needles, xi, 64
Cleve, Per Teodor, 107
Climate, 31–32
  humans influencing, 141
"Climate Change Action Plan, The," 262
Climate Change 1994, 235–36
Climate change reports, 235–36, 237–38
Climatic change, xiv, 31–32, 41–42,
  120–21, 122, 125–26, 128–29,
  133–34, 159–60, 196–98
  atmospheric gases in, 114
  civilizations doomed by, 129
  CO_2–induced, 224
  effect on plants and animals, 239–40
  effects of, 246–48
  human–induced, 262
  projected rates of, 208
  response of species to, 212–13
  scientists on, 199–201
  theories of, 164
  United Nations Conference on, 254–68
  and water supply, 241–43
Climatic flywheel, 203–9

Climatic history, 127, 217
Climatic pendulum, 159–71, 224
Climatologists, xi, 42, 162, 189, 201–2, 206, 231, 234
Clinton, Bill, 262, 263, 264, 267, 277
Clinton administration, 255, 262–63
Clouds, 201–03, 204, 206–7, 213, 252
Coal, 20–21, 29, 31, 35–36, 53, 56, 62, 68, 69, 167, 190, 259
  CO_2 produced by burning, 111
  consumption of, 142–44
  as fuel, 42, 43
  transporting, 71
Coal mining, 39–41, 43–44, 53
Coalbrookdale, xiii, 66, 68, 70–71, 168
Coalbrookdale Works, 70, 71
Coastal marshes, 245–46
Cockroaches, 28, 220
Coke, 68–69, 70
Cole, Henry, 75
Combustion, 32, 42
Concrete, 65
Condorcet, Marquis de, 7
Conservation, 143–44
Converter, 85–86, 87, 89, 90, 168
Cooling agent(s), 146–47
Copernicus, xii, 269
Corals, xiv, 213–14, 215, 228
Corbett, Robert, 61
Cormorants, flightless, 225
Corporations
  and global warming, 271
Cousteau, Jacques, 166
Couthon, Georges Auguste, 7–8
Crepuscularism, 14
Cretaceous period, 31
Crimean War, 84
Crompton, Samuel, 48–49
Cryptic moth, 14, 22–23, 53
Crystal Palace, 77–80
Curie, Marie, 105
Curie, Pierre, 105
Cushman, John H., Jr., 273
Cuvier, Georges, 25

Daimler, Gottlieb, 100–101
Daimler Motoren-Gesellschaft, 101
Darby, Abraham, xiii, 68–70
Darby, Abraham, II, 69–70
Darby, Abraham, III, 70
Darbys, 71, 80, 168
Darwin, Charles, xii, 15–18, 19, 20, 22, 23, 24, 26–27, 28, 53, 115, 174, 184, 185, 203, 207, 211, 225
da Silva, Darcy, 190–91
da Silva, Darly, 190
Davie, James, 25
Decay, 29, 30, 31
Deforestation, 181–84, 186, 189–91

Degas, Edgar, 63
Dekalb Company, 130–31
Dendrochronology, 119
Dengue fever, 226, 240–41
Descartes, René, 4, 8
Descent of Man and Selection in Relation to Sex, The (Darwin), 18
Destinies of the Stars, The (Arrhenius), 106
Des Voeux, H. A., 149–50
Detroit Automobile Company, 138
Developing countries, 242–43, 255, 263–64, 265, 267, 277
Dichlorodifluoromethane, 146
Dickens, Charles, 62, 181–82
Diffusion, 10–11
Diffusion equation, 10
Dinosaurs, 31–32
Disease, xv, 226–27
  temperature and, 247
  vector–borne, 240–41
Dotto, Lydia, 170
Doyle, Sir Arthur Conan, 149
Drake, Edwin Laurentine, 93–96, 137
Drought, xv, 129–30, 133, 196–97, 222, 228, 232–33, 249, 274–75
Dust/dust storms, 130, 188–89
Dust bowl, xiii, 130–31, 132, 133, 197, 233
"Dynamists," 218–19
"Dynamo and the Virgin, The" (Adams), 102

Earth
  origins and age of, xii, 24–26
  solar radiation, 162–63
Earth First!, 165, 270
Earth in the Balance (Gore), 260
École Polytechnique, 9, 10, 12
Edison Illumination Company, 138
Edith's checkerspot (butterfly), 212–13
Edleston, R. S., 13–14, 276
Ehrlich, Paul R., 164–65, 213
Eiffel Tower, 101
Eiseley, Loren, xi
Eizenstat, Stuart, 255, 256, 266, 277
Electricity, 34, 144, 258, 259
Electrolytic dissociation, 107, 109
Eliot, George, 72
Eliot, T. S., 42, 148–49
Ellsaesser, Hugh, 231
End of Nature, The (McKibben), 198
Enlightenment, 34
Entomologist (journal), 13, 14
Entropy (Rifkin), 166
Entropy Law, 166
Environment
  effect of human action on, 22, 183–84, 198, 276–77
  effect of industrialization on, 110–11

Environmental Defense Fund, 194, 195
Environmental groups, 195, 237, 256
Environmental Protection Agency (EPA), 245, 246, 251, 261, 274
Environmentalists, 189, 190–91, 202, 262
  and Kyoto conference, 257–60, 264, 265, 267
Ericson, David B., 163
Erik the Red, 122–23, 125–26, 161
Estrada, Raul, 266
European Union, 256, 266
Evolution, 15, 16, 19, 22, 184
Evolutionary Theory, 15–18
Experiments and Observations on Electricity (Franklin), 34
"Experiments in Plant Hybridization" (Mendel), 20
Extinction
  see Species extinction

Falcon Lake, 232–33
Family (Frazier), 182
Federal Emergency Management Agency, 223
Fiennes, Celia, 43
Filho, Francisco Mendes (Chico Mendes), 187–88, 190–91
Fire(s), xv, 222, 228, 233, 247–48, 270, 272, 274
  control, production of, 32–33
Firestone Tire, 140
Fish, 178, 179, 246, 253
Fitzroy, Robert, 26
Fixation, 30
Floderus, M. M., 108
Flooding, floods, 24–25, 29, 222, 223, 226, 249, 274, 275
Flux adjustment, 205–6, 208
Fogs, 148–50, 181
"Forcing," 207, 209
Ford, Henry, xiii, 23, 138–40, 143, 168, 274
Ford Motor Car Company, 138, 140
Foreman, Dave, 165
Forests
  saving and replanting of, 271–72
Forman, Joseph, xiv, 173–76, 189, 195
Fossil fuel industry, 261
Fossil fuels, xi, xiii, 31, 128, 154
  burning of, 251, 277
  consumption of, 142–44, 168, 198, 252
Fossil record, 17, 25, 28
Fouquier–Tinville, Antoine, 7
Four Corners, 116, 117, 119, 120, 121, 122, 125
Fourier, Jean–Baptiste–Joseph, xii, xiii, xv, 3–12, 15, 25, 33, 109, 113
Fourier series, 10
Fox, William Darwin, 16

Framework Convention on Climate Change, 262
Franklin, Benjamin, 34–36, 55, 112
Franklin (Philadelphia) stove, 34–35
Frazier, Ian, 182
Freons, 147
Frick, Henry Clay, 90–91
Friends of the Earth, 257, 259–60, 264
Frigidaire, 146
Frogs/toads, dying, 276–77
Frost, Robert, 51

Galápagos Islands, 213, 225, 228
Galileo, xii
Gas engine, 99–101
Gaskell, Elizabeth, 20
Gasmotorenfabrik Deutz, 100
Gasoline, 99, 145, 258, 259
General Circulation Models (GCMs), 205, 206, 207, 250
General Motors, 140
General Motors Research Corporation (Delco), 145
"General Remarks on the Temperature of the Terrestrial Globe and Planetary Spaces" (Fourier), 11–12
Genetics, 19–20
Geologic cycles, 111
Geologic eras, 28
Geologic time, 26, 27–28
Geology, xii, 25
Geophysical Fluid Dynamics Laboratory (GFDL), 204
George C. Marshall Institute, 200–201
Gershwin, George, 134
Giotto, 64
Glaciers, xi, xiv, 41, 142, 159, 161, 162, 163, 218
  melting, shrinking, 194, 203
Glasgow, 58, 61, 179
Glass, 77–78
Global Climate Coalition, 258, 260, 268
Global warming, xi–xii, xiii, xiv–xv, 3, 12, 231–34, 269–77
  as benevolent phenomenon, 115, 249, 251–52
  debates over, xiii–xiv, 253
  effect on corals, 214
  evidence of, 213, 215, 277
  forecast of, 207, 108
  human activity in, 110, 167–68, 194
  ice sheet and, 218–19
  Sagan on, 210–11
  scenarios, 235–53
  world's leading authority on, 235
Global warming theory, 171, 237
  critics of, 244–51
Global warming treaty, xv, 277

Gore, Al, 255, 259, 260–61, 262–64, 265, 267, 268, 275, 276, 277
Gould, Jay, 98
Gradualism, 17, 18
Grateful Dead, 190
Great Depression, xiii, 91, 132
Great Drought, 120–21, 131
Great Exhibition of the Works of Industry of All Nations, 75–80
Great Exposition (Paris), xiii, 102
Great Plains, 129–30, 159, 196–97
Great Smog (London), 151, 227
Great Western Railway, 81
Greenfield Village, 139
Greenhouse effect, xii, xiv, 33, 44–45, 114, 128–29, 132, 137, 142, 148, 159, 181
  CFCs in, 170
  critics of, 237
  El Niño and, 223–24
  evidence of, 196–98
  human–induced global warming engine of, 167
  misnomer, 210, 211
  secondary, 189
"Greenhouse fingerprint," 208
Greenhouse gas emissions
  limits on, 262, 266–67, 273
  lowering, 256, 262, 263, 275–76
Greenhouse gases, xv, 147, 170, 190, 200, 202, 203, 204, 207, 208, 209, 219–21, 238
  countries responsible for release of, 235
  effect on climate, 259
  and global warming, 231
  increase in, 218, 221
Greenhouse research
  models in, 200, 204–9
Greenland, 122, 123–28, 129, 161, 171, 203, 249
  Eastern Settlement, 123–24, 127–28
  Summit Camp, 216–17
  Western Settlement, 122, 124, 125, 126, 127
Greenland Ice–Core Project (GRIP), 127
Greenlander's Saga, 123
Greenpeace, 166, 257, 262
Groves, Leslie, 254
Guillotin, Joseph Ignace, 6
Guillotine, 6–8
Gunn, J. M., 171
Gwynne, Peter, 164

Hagel, Chuck, 263, 264, 268
Hall, Julia ("Butterfly"), 270–71
Hall–in–the–Wood, 49
Halley Bay, 173, 174
Halocarbons, 266

Hansen, James E., 196–98, 199, 200, 231–32, 234, 235, 238, 261, 275
Hard Times (Dickens), 62
Hargreaves, James, 46, 47, 49, 50
Hashimoto, Ryutaro, 266
Heat, xv, 132–33, 196–98, 247, 270
  from Sun, 11–12
  and water supply, 242
Hemingway, Ernest, 87–88
Henn, Rainer, 192–93
Henry Bessemer and Company Ltd., 87
Henry VIII, 44, 67
Henslow, John Stephens, 26
HMS *Beagle*, 15, 26, 211
Hoar, Timothy J., 224
Högbom, Arvid, 111, 112
Homestead (Pa.) strike, 90–91
Hooker telescope, 135–36
Hot Talk, Cold Science (Singer), 249, 273
Hothouse, 113–14
House of Commons
  Select Committee on the Smoke Nuisance, 57
Howard, Luke, 201
Hubble, Edwin Power, 135–36
Hubble, Grace, 166
Hugh More fund, 164
Human action
  alteration of species, 23, 24
  and climate, xiv, 141, 163, 209
  effect on environment, 22, 183–84, 198, 276–77
  effect on seas, 166
  and global warming, 110, 167–68, 194
  and greenhouse gases, 219–20
Humboldt, Alexander von, 26
Huon pines, 215
Hurricanes, 244–45, 275
Hutton, James, xii, 25–26, 27, 28, 29, 269
Hydrological cycle, 202
Hydropower, 241

Ice Age, xi
Ice ages, 111, 162, 163, 171
Ice cores, xiv, 127, 128, 167, 171, 217, 250
Ice sheets, xi, 218–19, 238
Ice shelf, 217–18
Icebergs, 215, 217
Iceland, 122, 123, 125, 252
Iceman, 192–94
Illustrated London News, 177
India, 167, 190, 219, 243, 264, 277
Industrial age, xiii, 29, 34, 41, 44, 80, 102, 277
  energy consumption, 166
Industrial chimneys, 55–65
Industrial lobbyists, 273
  Kyoto conference, 257–60, 263, 265, 267–68

Industrial melanism, 23
Industrial pollutants/pollution, xiii,
    112–13, 177–79, 215
Industrial revolution, xi, 12, 27, 32, 59,
    70, 92–93, 168, 219, 273, 276
  energy sources, 96
  power loom in, 51
  steam engine in, 53
Industrialists, 98–99
Industrialization, 110–11, 141
Infrared radiation, 33, 114, 132, 137, 202,
    203
  CFCs trap, 170
  effect of clouds on, 206
Ingenhousz, Jan, 34
Inheritance, 16, 17, 19
Inland Voyage, An (Stevenson), 54
Insects, 28, 240, 249, 251
Internal combustion engine, 99,
    100–101, 137, 144
Intergovernmental Panel on Climate
    Change (IPCC), 235–36, 237, 238,
    242, 244, 251, 263
International Conference on the Chang-
    ing Atmosphere, 235
International Geophysical Year (IGY),
    154, 155, 156, 173
International Meteorological Commis-
    sion, 201
Introduction to Weather and Climate, An
    (Trewartha), 132
Inuit/Eskimo, 124, 126
Inversion, 137
Iron, 65, 66–71, 77–78, 80, 144
Iron Bridge, 70–71
Iron–frame building(s), 78
Iron ore, 86–87
Island states, 243–44, 256
Italianate style, 63–64

J. C. Gostling and Company, 59–60
J. Edgar Thomson Steel Works, 89–90
Jack the Ripper, 149
Jacobs, Stanley, 217
James I, king of England, 43
Jastrow, Robert, 200
Java Sea, 160
Jefferson, Thomas, 159
Jenkin, Fleeming, 16
Jennings, Peter, 258
Jockeybush Lake, 179
Johannessen, Ola M., 218
John A. Roebling's Sons, 133
Johnson, Andrew, 83
Johnston, J. Bennett, 196
Jones, Rhys, 248–49
Jordan, Thomas, 39–40
Jurassic period, 31

Kareiva, Peter, 213
Kay, John, 45–46, 50
Keeling, Charles, xiv, 151–57, 166–68,
    173, 174, 175, 260
"Keeling curve," 167, 171, 236
Keilholtz, Lester S., 145–46
Kelly, William, 87
Kennert, James, 220–21
Kenya, 226–27
Kin Klethla, 116, 121
Kinkel, Klaus, 255–56
Kipling, Rudyard, 163
Kiribati, 243–44
Kirtland's warbler, 239
Kittlewell, Bernard, 18
Knights of St. John, 10
Kohl, Helmut, 266
Kohlrausch, Friedrich, 109
Krakatau eruption, 111–13, 160
Krieder, Kalee, 257
Kristof, Nicholas, 243
Kuala Lumpur, 227
Kyoto, Japan, xv, 254–55
Kyoto conference/summit, 254–68, 269,
    271, 272, 273, 277
Kyoto Protocol, 266–68, 273, 274, 275–76

Ladybird beetles, 14
Lagrange, Joseph–Louis, 8, 9, 10, 11
Laki (volcano), 112
Lamarck, Jean–Baptiste, 201
Landscape with Smokestacks (artwork), 63
Langen, Eugen, 100
Langley, Samuel P., 110
Laplace, Pierre–Simon, 8, 9, 10, 11
"Large Losses of Total Ozone in Antarctica
    Reveal Seasonal ClO_x/NO_x Interac-
    tion" (Rowland and Molina), 175
Lavoisier, Antoine, 7
Lead tetraethyl, 145
LeBegue, Edmie Germaine, 4
Lenoir, Étienne, 99
Lief the Lucky, 122
Life in the Universe (Arrhenius), 106
Limestone, 67, 111
Lincoln, Abraham, 183
Lindzen, Richard, 199–201, 202, 209, 252
Linnaeus, Carolus, 21
Little America, 156
Little Climatic Optimum, 120, 161
Little Ice Age, 41–42, 122, 125, 161, 250
Liverpool and Manchester Railway, 72
Livestock raising, 220
Lockheed Martin (co.), 271
Locomotive, 72–74
London, 35–36, 42, 43, 133
  fogs, 148–50
  Great Smog, 151, 227

smogs, 150–51
London, Edinburgh, and Dublin Philosophical Magazine and Journal of Science, 109, 111
London International Exhibition, 87
Loom(s), 45, 50–51, 52
Los Angeles, 135, 136, 137, 140
Los Angeles Air Pollution Foundation, 154
Los Angeles basin, 137, 150
Lott, Trent, 268
Louisette/Louison (guillotine), 6–7
Love Canal, 261
"Love Song of J. Alfred Prufrock, The" (Eliot), 149
Lowell, James Russell, 148
Lyell, Sir Charles, xii, 26–28, 29, 238, 269

McDonald, Gail, 260
McKibben, Bill, 198–99
Malaria, 226, 240
Malay Archipelago, 184
Malay Archipelago, The (Wallace), 185
Mammoth, 158–59
Man and Nature (Marsh), 183
Manabe, Sukuro, 208–9
Manchester, England, 13, 14, 18, 20–21, 22, 23, 177, 276
Manometers, 152, 155, 156–57
Marsh, George Perkins, 183–84, 188, 198
Mason, Charles, 117
Mauna Loa, 156
Maybach, Wilhelm, 100, 101
Meacher, Michael, 256
Mechanical technology, 138
Medieval Warm Period, 120, 215
Meehl, Gerald A., 224
Mehra, Malini, 257
Melville, Herman, 82–83
Memoires de l'Académie Royale des Sciences, 12
Mendel, Gregor, 19–20, 23
Mesa Verde, 117
Methane (CH_4), 170, 203, 219–21, 266
Methyl chloride, 146
Mexico, 247–48
Middlemarch (Eliot), 72
Midgley, Thomas, Jr., xiv, 145–47, 168, 170, 176
Mies van der Rohe, Ludwig, 78
Milankovitch, Milutan, 162–63
Milankovitch cycles, 163–64, 168, 204
Miles, Martin W., 218
Milky Way, xi, 136
Mill, John Stuart, 107
Model T, 138–39, 140
Models/modeling, 200, 201–9
Molina, Mario J., 168–69, 170, 174, 175, 195

Monge, Gaspard, 9, 10, 12
Montreal Protocol, 195–96
Moore, Thomas Gale, 253
Morgan, John Pierpont, 98
Moths, xii, 13–14, 18, 22–23
Mount Palomar Observatory, 136
Mount Pinatubo, 234
Mount Tambora, 160
Mudslides, xv, 223
Mushet, Robert F., 87

N.A. Otto and Company, 100
Namias, Jerome, 200
Napoleon, Prince, 84
Napoleon Bonaparte, 3, 9, 10, 11, 64, 71, 72
NASA, 161, 176
National Academy of Sciences, 175, 176, 273
National Center for Atmospheric Research (NCAR), 204, 205, 208, 248
National City Lines, 140
National Convention (France), 7, 8
National Geographic Society, 166
National Ice Core Laboratory, 127
National Oceanic and Atmospheric Administration (NOAA), 228, 234
National Science Foundation, 216, 217
Natural Environmental Research Council, 174
Natural History Society of Brünn, 20
Natural philosophy, 8, 166
Natural Resources Defense Council, 195, 265
Natural Science Society, 19
Natural selection, 15, 16, 19
Nature (journal), 168, 171, 175, 195, 212, 224, 237
Neanderthals, 32
Negative feedback, 202, 215, 252
Nelson, Horatio, 9–10
Neolithic period, 128
Neolithic revolution, 198
New York City, 133–34, 197
New York Steam Heating Company, 62
New York Times, 133, 272
New Yorker, 198
Newcomen, Thomas, xiii, 44, 52, 70
Newell, Reginald, 200
Newland's Mills, 59
Newsweek, 164
Newton, Isaac, xii, 3, 5, 21, 34, 51, 162, 166, 174
  law of cooling, 110
  law of gravitation, 8
Nierenberg, William A., 200
Nimbus 7 (satellite), 176
Niña, La, 223, 274–75

Nineteen Eighty–four (Orwell), 62
Niño, El, xv, 214, 222–31, 234, 240, 245, 250, 253, 269, 272, 274, 275
    template for global warming, 248
Nitric acid, 177
Nitrogen oxide, 177, 266
Noachian Deluge, 25, 197
Nobel, Alfred Bernhard, 105
Nobel prize, 105, 107, 109
Norse, 122–28
Northern Hemisphere, 141, 157, 171, 176, 203, 215, 238, 240, 243
Norway, 178, 179
Nuclear power, 241–42

Observations on the Causes and Cure of Smoky Chimneys (Franklin), 34–35
Ocean eddies, 208
Ocean temperature, 220–21
    rise in, 214, 222, 223, 229–30
Oceans, 217–18
    methane in, 220–21
    sinks for CO_2, 141–42, 154, 168, 189, 250
Oden, Svante, 178, 179
Oil, 31, 94–99, 137–38, 144, 168
    consumption of, 143
O'Keefe, William F., 258, 267–68
Oki, Hiroshi, 256
"Okies," xiii, 131, 132, 140
Oklahoma Panhandle, 233
Olby, Robert C., 19
Olofsson, Lasse, 107
"On the Influence of Carbonic Acid in the Air upon the Temperature of the Ground" (Arrhenius), 109–10
On the Origin of Species by Means of Natural Selection (Darwin), 15, 16, 18, 19, 22
Oppenheimer, Michael, 194, 271–72
Oregon Institute of Science and Medicine, 273
Orwell, George, 62
Oscar II, king of Sweden, 105
Ostwald, Friedrich Wilhelm, 106, 108–9
Otto, Nikolaus August, 99–100
Otto Silent Gas Engine, 100
Ozone, 169, 170, 180, 190
    effects of, 247
"Ozone: The Crisis That Wasn't" (Farman), 176
Ozone hole, 175, 176, 189, 194, 261, 276–77
Ozone layer, xiv, 147, 169
    depletion, 174–76, 195–96
"Ozone War," 170

Pacific Electric Railway Company, 140
Paleozoic, 28–29

Palu, 111, 112
Pangenesis, 16, 19
Paris exhibition (1867), 99
Park, Robert L., 273–74
Parmesan, Camille, 211–13
Pascal, Blaise, 3, 5
Patents, 45, 46, 48, 50, 52, 53, 68, 72, 83, 84, 147
Paul Bunyan Comes West, 184
Paxton, Joseph, 76–77, 78, 79–80
Peas, 19, 20, 207
Peat, 29, 31
Pelletier, Nicolas Jacques, 6
Pemberton, Henry, 34
Penguins, 225
Pennines, 22, 276
Pennsylvania
    Oil Creek, 94, 95
Pennsylvania Railroad Company, 88–89
Pennsylvania Rock Oil Company, 94
Peppered moth, 14, 18, 22–23, 276
Percival, Dr., 53
Personal Narrative (Humboldt), 26
Peru, 222–23
Pettersson, Otto, 108
Pew Center for Global Climate Change, 271
pH levels, 177, 178, 179
Photosphere, 160
Photosynthesis, 30, 32, 157, 214
Pig iron, 68, 69, 85
Pitlochry, Scotland, 176–77
Planet modeling, 204–9
Pleistocene, xi, 41, 158, 159
Plutarch, 94
Polar ice, 252
    melting of, 202, 243
Pollack, Henry, 277
Pollutants
    in acid rain, 177
Pollution, 20, 21, 23, 42, 44, 62, 166, 227–28
Population bomb, 164–65
Port Dundas Chemical Works, 58, 61
Positive feedback, 202, 207, 214–15
Precambrian, 28, 31
Precession of the equinoxes, 162, 163
Principia (Newton), 21, 34
Principles of Geology (Lyell), 26–27
Prize Essay, 11
Proceedings (Natural History Society of Brünn), 20
Ptolemy
    epicycles, 206
Pullman cars, 88–89

Rae, Michael, 257
Rail transport/railroads, 71–74, 77, 81, 98, 140

Rain forests, xiv, 185, 186–89, 190–91, 228, 247, 248, 272
Rains, xv, 129–30, 197, 225
Ramanathan, Veerabhadran, 170
Ramses II, 64
Reagan, Ronald, 100
Reagan administration, 194
Refrigerants, 146–47
Reilly, William, 261
Revelle, Roger, 155, 168, 260
Rice, 219–20
Rifkin, Jeremy, 166
Rift Valley Fever, 226
Robespierre, Maximilien, 7, 8, 9
Rockefeller, John D., xiii, 96, 97–99, 101, 168, 274
Rockefeller, William, 97
Roosevelt, Teddy, 143–44
Ross Sea, 217
Rowland, F. Sherwood, 168–69, 170, 174, 175, 176, 195–96
Royal Meteorological Society, 148
Royal Society of Arts, 75
Royal Society of London, 12
Royal Swedish Academy of Sciences, 105, 109
Ruskin, John, 63

Sagan, Carl, 210–11, 221
Saint–Just, Louis de, 7, 8
St. Rollox Chemical Works, 58, 63
Salinger, M. J., 171
Salinity, 242, 243
Saltation, 16, 17
Santer, Benjamin, 208, 209
Sarawak, 227
Sarney, Jose, 190
Satellite data, 161, 237–38, 250
Scandinavia, 178
Scenarios, 235–53
Schiff, Harold, 170
Schlesinger, Michael, 199
Schnabel, Matthias, 237–38, 250
Schneider, Steve, 201
Science (journal), 170, 171, 199
Science and Environment Policy Project, 273
Science Digest, 176
Science News, 197
"Scientific Perspectives on the Green-house Problem" (Nierenberg, Jas-trow, and Seitz), 200–201
Scientists
  on global warming, 164, 199–201, 202, 236, 259, 261, 269, 271, 273–74, 275
Scott, Robert Falcon, 173
Scott, Sir Walter, 49
Scott, Thomas A., 88

Scripps Institution of Oceanography, 155, 166
Scrope, George, 27
Sea level, rise in, 203, 208, 217–18, 243–44, 245–46, 256
Seagram Building, 78
Seas
  CO_2 stored in, 33
  warmer temperatures in, 252–53
  see also Oceans
Seitz, Frederick, 200, 273
Seneca Oil Company, 93
Shackleton, Ernest Henry, 173
Shelley, Percy Bysshe, 56–57
Shells, 84–86
Shuttle Club, 45
Sibthrop, Charles, 77
Signal Corps, 110
Singer, S. Fred, 249–50, 251, 252, 273
Sizer, Nigel, 272
Skrälings, 124, 126
Slade, Tuiloma Neroni, 256
Slag, 67, 85, 89
Smith, Robert Angus, 177
Smith, William "Uncle Billy," 94–95
Smith, William "Strata," 27
Smithsonian Institution, 183
Smogs, xiv, 136–37, 150–51, 227
Smoke, 35, 53, 55
Smoke Abatement League of Great Britain
  Manchester Conference, 149–50
Social Darwinism, 98
Solar Max (satellite), 161–62
Somalia, 226
Sommerville, Richard J., 206
South America, 183
  El Niño in, 225–26
South Pole, 175–76
Southern Oscillations, 224
Space shuttle, 170
Species
  alteration of, by humans, 23, 24
Species creation, 207
Species extinction, 28, 212, 238–40, 245, 246, 253, 276–77
Spectrophotometer, 173, 174–75, 176
Spindler, Konrad, 193
Spinning jenny, 46, 47, 48, 49
Sport utility vehicles, 274
Spradley, J. R., 260
"Stabilists," 218–29
Standard Oil Company, 96, 98
Standard Oil of California, 140
Stanton Iron Works Company, 61–62
Steam engine, xiii, 44, 51, 52–53, 55–56, 71–72, 99, 102
Steamships, 81–83
Steel, 78, 84–87, 89–91

Stephenson, George, xiii, 72–74
Stephenson, Robert, 73–74
Stevens, William K., 213, 234
Stevenson, Robert Louis, 54–55, 63
Stimson, Henry L., 254, 255
Storms, 222, 223, 229, 230–31
Stratosphere, 147, 170, 211
    antarctic, 175, 176
    CFCs in, 169
Suess, Hans, 155, 168
Sulfur dioxide, 146, 150, 177
Sulfuric acid, 160, 177
Sun, 170
    and global temperature fluctuations,
    160–63
Sunoco (co.), 271
Sununu, John, 200
Supersonic Transport (SST), 170
Survival of the fittest, 22
Sweden, 178, 179
Systema naturae (Linnaeus), 21

Tall Chimney Construction (Bancroft and
    Bancroft), 55
"Tall chimneys," 55, 57–65, 168, 177,
    178–79
Tamman, Gustav, 113, 114, 115
Taylor, Zachary, 183
Technological change, 102
Tellus (journal), 155
Temperature change/fluctuation, 122,
    125–26, 128, 129, 141, 148, 159–71
    benefits of, 115, 142
    effects of, 246–48, 251
Temperature records, accuracy of, 250
Temperatures, xiii, 31, 41–42, 120, 235
    arctic, 216
    effect of CO_2 on, 113–15
    high, 196, 197
    measurements, 237–38
    rise in, 33–34, 141, 148, 159–61, 207,
    208, 209, 211, 218–19, 231–34,
    246–48, 251, 275, 277
    simulation of, 206
    and tree ring growth, 215
    water, 246, 252–53
    see also Ocean temperature
Termites, 220
Terror (France), 6–7
Textile production, 44–52
Thames, tunnel under, 80–81
Theory of the Earth (Hutton), 26, 27
Thermidorian reaction, 9
Thermodynamics, first and second laws
    of, 166
Thomas, Lee M., 194
Thomas, Sidney Gilchrist, 87
3M (co.), 271
Thutmose III, 64

Tierra del Fuego, 175
Time, xii, 26
TOMS (Total Ozone Mapping Spectrom-
    eter), 176
Toolik Lake, 216, 218
Tornado Alley, 230–31
Törnebladh, H. R., 105
Townsend, Joseph, 58–59, 61
Toynbee, Arnold, 162
Travels on the Amazon and Rio Negro
    (Wallace), 184
Tree death, 179–81
Tree ring growth, xiv, 119, 120, 121, 215,
    250
Trees
    moving with climate change, 239–40
Trenberth, Kevin E., 205, 208, 224, 248
Trevithick, Richard, xiii, 71–72
Trewartha, Glen Thomas, 132–33
Tropical Forest Action Plan, 189
Troposphere, 137, 147
Truman, Harry, 254
Trumka, Richard, L., 265
Tsunamis, 112
Tutt, J. W., 22
Twain, Mark, 91
Tyndall, John, 110
Typhoon Winnie, 233

Ulfsson, Gunnbjörn, 122
Uniformitarianism, xii, 26, 27
United Nations, 189, 235
United Nations Conference on Climate
    Change, xiv, 254–68
United Nations Conference on Environ-
    ment and Development (first Earth
    Summit), 262
United States
    environmental treaties, 262
    and Kyoto conference, 255–56,
    263–67, 277
    limits on CFCs, 195
United Technologies (co.), 271
Uppsala, Sweden, 110–11
Urban heat island effect, 141, 250
U.S. Clean Air Act of 1970, 178–79
U.S. Greenland Ice–Sheet Project 2
    (GISP2), 127
U.S. Senate, xv, 268
    Committee on Energy and Natural Re-
    sources, 196
U.S. War Department, 110
U.S. Weather Service, 154, 156, 157
Ussher, James, 24

Vanderbilt, Cornelius, 98
Van't Hoff, Jacobus H., 109
Vector–borne diseases, 240–41
Vertical integration, 98

Victoria, Queen, 75, 78, 79, 83, 250
Victorian age, 55
Vikings, xiii, 122, 128, 129, 161, 249
Vinland, 123
Virgin Islands, 213
Volcanic ash, 112
Volcanism, 26
Volcanoes, 24–25, 160, 163, 168
Voyage of the Beagle, The (Darwin) 28

Waldsterben, 180
Wallace, Alfred Russel, 15–16, 184–86
Washington, Warren M., 224
Water supply, 241–43
Water vapor, 114, 149, 170, 177, 202, 208,
    218, 219, 221, 252
Waters
    CO_2 in, 30–31
    see also Oceans; Seas
Watt, James, xiii, 52–53, 55, 71, 72
Weather, extreme, 208, 224, 225–31
Weather cycles
    cooling and warming, 159–71
    drought–cool, 129
    dry–wet, 119–20
Weaving machine, 50–51
Webb, James, 253
Weber Company, 65
Weiner, Jonathan, 162

Wentz, Frank J., 237–38, 250
"Wet deserts," 186–87
Wetlands, 245
Wetherill, Al, 117
Wetherill, Richard, 117
Whale oil, 92–93, 94
Whaling industry, 92–93
Whirlpool (co.), 271
Wilderness Society, 165
Wilson, Alex T., 182–83
Wilson, Charles Erin, 274
Wollin, Goesta, 163
Wood, 42
    consumption of, 35
Woodruff Company, 89
Wordie shelf, 218
World Bank, 191, 272
World Health Organization, 227, 240
World Resources Institute, 195
World War I, 144
World Wildlife Fund, 267, 272
Worlds in the Making (Arrhenius), 106,
    112–13, 115

Yeltsin, Boris, 265
Yokich, Stephen, 263
Younger, Dryas, 128

Zoologist (journal), 13